Medical Error and Patient Safety

Human Factors in Medicine

Medical Error and Patient Safety

Human Factors in Medicine

George A. Peters

Barbara J. Peters

CRC Press
Taylor & Francis Group
Boca Raton   London   New York

CRC Press is an imprint of the
Taylor & Francis Group, an informa business

CRC Press
Taylor & Francis Group
6000 Broken Sound Parkway NW, Suite 300
Boca Raton, FL 33487-2742

Library of Congress Cataloging-in-Publication Data

Peters, George A., 1924-
    Medical error and patient safety : human factors in medicine / George A. Peters and Barbara J. Peters.
        p. ; cm.
    "A CRC title."
    Companion v. to: Human error. 2006.
    Includes bibliographical references and index.
    ISBN-13: 978-1-4200-6478-0 (hardcover : alk. paper)
    ISBN-10: 1-4200-6478-9 (hardcover : alk. paper)
    1. Medical errors--Prevention. 2. Patients--Safety measures. I. Peters, Barbara J., 1950- II. Peters, George A., 1924- Human error. III. Title.
    [DNLM: 1. Medical Errors--prevention & control. 2. Safety Management. WB 100 P481m 2008]

R729.8.P44 2008
610.28'9--dc22   2007024833

Visit the Taylor & Francis Web site at
http://www.taylorandfrancis.com

and the CRC Press Web site at
http://www.crcpress.com

Contents

Preface

The medical error problem has been recognized as something very serious in terms of avoidable patient injury, achieving efficacious treatment, and controlling health care costs. The prevention of medical errors may seem to be a relatively simple task, and with heightened awareness, some improvements have been reported. However, the search for reasonable, acceptable, and more effective remedies and countermeasures continues with ever-greater vigor. Medical error has proved to be a difficult and recalcitrant phenomenon.

The contents of this book provide a source of useful ideas, concepts, and techniques that could be selectively applied to reduce an intolerable rate of unacceptable errors, mistakes, goofs, or shortcomings in expected human performance. With better understanding, knowledge, and directed motivation, there should be rapid advancement in the medical error management discipline.

This is a companion book for the treatise *Human Error: Causes and Control* published in 2006 by CRC/Taylor and Francis Group. That book provides supplemental information and insight that may be useful in the task of appropriately controlling the causes of unwanted error.

Since medical error is situation-specific and involves diverse human performance, the choice of appropriate preventive and corrective techniques is critical.

This book is written to be comprehensible and instructive for students, thought provoking and useful for expert specialists, and interesting and stimulating for those involved in research. The contents present many ideas within a fairly broad perspective, so the reader may pick the concepts or cherries that seem ripe and appropriate, discard those that seem overripe or not appropriate to a particular situation, and return later when the unripe ideas have matured in the personal experience of the reader and may then meet the reader's needs. There may be strong arguments against some concepts or applications, as may be expected, but such disagreements should stimulate research that will help in medical error management practices.

The methodology utilized may produce differing results based on exactly how it is applied. A preventive program requires considerable professional discretion, social sensitivity, and compliance with ethical responsibilities and constraints. We have attempted to provide useful information, but it is the reader's responsibility to fully comprehend the concepts, correctly identify the needs, and undertake appropriate actions. That effort is well worthwhile since success can be measured in lives saved, injuries avoided, health care improvements, and cost reductions.

Note: We have attempted to maintain openness and independence of thought during the preparation of this book. Elimination of possible bias was achieved by not accepting any financial support, grants, funding, or direct assistance from those who might have a vested interest in the outcome of this venture. There was no third-party direction or control. Independence was also maintained by presenting all significant sides of contested issues or topics.

G. A. Peters
B. J. Peters
Santa Monica, California

The Authors

George Peters is a multidisciplinary specialist who has been a frequent contributor to legal, engineering, scientific, and medical journals as an author, columnist, editor, or advisor. Early in his career, he worked at a major general hospital in a behavioral research and diagnostic capacity. Subsequently, Dr. Peters had experience as a design engineer, safety specialist, reliability engineer, and quality engineer. He has served as a director on several certification boards, held elective offices  in several professional societies, participated on many standards formulation committees, and has received many awards. He is a former fellow of the Royal Society of Health (United Kingdom) and has been elected as a Fellow of the American Association for the Advancement of Science.

Barbara Peters has specialized in problem solving relating to medical error, safety, risk assessment, and environmental health hazards. She has technically evaluated medical devices, performed detailed reviews and analyses of medical records and procedures, observed medical protocols and customary practices, and conducted extensive interviews with medical specialists, health care technicians, and patients. Dr. Peters is a fellow of the Roscoe Pound Foundation and,  with her father, George A. Peters, has authored or served as editor of 41 professional-level books.

1

Introduction

Common Understanding

Since the dawn of civilization, people were aware that they could inadvertently commit errors or mistakes. With modern industrialization, conscious attempts were made to reduce worker error or fault. There was an awareness that medical personnel could make mistakes, but there was a great deal of tolerance because of the noble objectives of learned professionals in the health care industry.

Many publications concerning human error, generally describing the incidents in which it occurred and the damage that resulted, were available. If prevention was considered, then it was usually a general admonition or advisory to stop such behavior. The attitude was that mistakes happen, and committing error was a normal human trait. Signs were posted telling people to pay attention, act in a safe manner, and be responsible. Errors were the fault of the person committing them, not the machines they operated, the procedures that they were given, or the environment in which they worked or lived.

Sophisticated Knowledge

The advent of expensive and complex space and missile systems in the early 1960s resulted in a much greater focus on human error because of the likelihood of catastrophic consequences. We routinely published or provided technical guidance on error during that time and subsequently (Peters et al. 1962; Peters 1963, 1964, 1996; Peters and Peters 2000, 2006a). Human error prevention concepts were applied to health problems for which there were high error frequency rates, high injury rates, and high-severity consequences. As indicated in the chapters of this book, specific techniques on error prevention were developed and applied in various industries and professions.

What transpired during the 1960–1999 time period was the growth of risk assessment procedures, evolution of causation techniques, and the development of various safety criteria regarding which risks were acceptable. If a little benefit was to be derived from error reduction, then the status quo prevailed, nothing was done, no costs were incurred, and complacency was prevalent. If some benefit was to be derived from error reduction, then something routine might be deemed necessary, such as compliance with trade standards and following the custom and practice of the industry or occupation. If there were appreciable adverse consequences from errors, then a strong zero-error-type program might be implemented (see chapter 5, this volume, on analysis).

Current Urgency

In 1999, it came as a surprise when a federal agency, the Institute of Medicine (IOM), indicated, in the now-famous report *To Err Is Human* (Kohn et al. 1999), that between 44,000 and 98,000 Americans die each year from medical errors. Within weeks, the U.S. Congress held hearings on medical errors and patient safety, and the general public became aware of the serious health problems that could affect anyone at any time. Physicians, nurses, and hospitals became alerted, motivated, and active in attempting to improve the quality and safety of health care. There were some favorable results, but most commentators believed much more was needed., and despite many approaches, studies, publications, discussions, and activities between 1999 and 2006, only incremental advances in knowledge resulted. There was no magic remedy, only a seemingly complex and intractable human behavior problem. Some of the complexity seemed to be man-made. The current status of the medical error problem is suggested by the following discussion.

Medication Errors

Seven years after the 1999 IOM report brought attention to the problem, a July 2006 IOM report stated that medication errors harm 1.5 million people each year in the United States. Medical errors kill several thousand people a year. Hospital patients should expect to suffer one such error each day of hospitalization. Such error rates vary among hospitals, and not all errors are serious (Anderson 2006).

Pediatric Dosing Errors

Of new prescriptions written for children during outpatient visits to health maintenance organizations (HMOs), about 15% were for the wrong dose. About 8% were potential overdoses, and 7% were underdoses. Analgesics (pain relievers) were most likely to be overdosed (15%), and antiepileptics were most likely to be underdosed (20%). Error rates were no lower for electronic prescription writers than with the use of paper prescriptions (McPhillips 2005).

Prescription Drug Listing

A Drug Registration and Listing System was mandated by the 1972 Drug Listing Act. The U.S. Food and Drug Administration was to compile a list of all prescription drugs manufactured, prepared, or compounded for commerce. As of 2005, some 33 years later, an estimated 9,187 prescription drugs were missing from the list, and another 5,150 drugs were not listed because of errors. The inspector general of the Department of Health and Human Services blamed the Food and Drug Administration (FDA) for failure to set up an efficient listing process and a lack of oversight (CEN 2006).

Toxic Exposures

When a possible toxic exposure exists, the treatment may vary by the extent of the exposure, the symptoms manifested, and the chemical and its toxic effects. However, 30,000 chemicals that have not been tested regarding toxicity or had appropriate risk impact studies are in use, so there is no objective foundation for verifiable opinions (Peters and Peters 2006a). The European Union has a REACH directive that calls for the registration, evaluation, authorization, and possible restrictions on such chemicals. If fully enacted, the results would not become available for several years.

The European Union REACH regulations had been debated for many years and were opposed by a number of countries as contrary to World Trade Organization principles because they were overly burdensome and difficult to implement. However, acceptance of a weakened version, by Europe's Council of Ministers, occurred at the end of 2006 ("EU Regs" 2006). The regulations, effective June 1, 2007, requires manufacturers to test industrial chemicals used in the manufacturing process and to gather health and safety data (Jacoby 2006). For the 1,500 chemicals deemed most dangerous, a reauthorization every 5 years might be required. Such information would be helpful to medical specialists dealing with toxic exposures.

A somewhat similar burden exists for toxic substances for which there is a permissible exposure level, threshold limit value, oncosafe level, or about a dozen other types of exposure levels (Peters and Peters 1980; Naim and Lemesch 1993). The problem is that major changes are almost continuously proposed, usually in the form of lower exposure limits, and information gradually accumulates regarding harmful effects of single or combined chemicals. New chemical compounds abound for which there is little toxicological information.

Laboratory Errors

Serious medical mistakes may start with error-prone systems for collecting, labeling, and handling blood samples and tissue specimens. Error rates of 3% to 5% have been reported (Landro 2006b). Blood may not be drawn correctly, or another patient's blood may be mistakenly used for analysis. A biopsy for a cervical cancer diagnostic purpose may miss the transformation zone of the cervix and fail to discover a lesion 30% to 40% of the time, resulting in a false negative. Fine-needle aspiration from thyroid nodules to rule out cancer may miss tumors 25% of the time. A false negative or false positive from a pathology laboratory may initiate serious treatment options and could have severe patient reactions. In malpractice claims that were pathology related, 63% involved false-negative and 22% false-positive cancer diagnoses.

Patient Identification Errors

An average of 26% of the babies in a neonatal intensive care unit were found to be at risk for being mistaken for another baby in the same unit on any given day. The confusion was that babies may share last names (34%), similar sounding surnames (9.7%), or have similar medical record numbers (44%); therefore, they were considered at risk. In addition, clinicians rely on standardized wristbands for patient identification, and these may have errors of content or be missing (Gray 2006). *At risk* means the possibility of receiving the wrong medication or treatment and assignment to the wrong mother.

Adverse Event Reports

For medical devices, adverse reaction reports and device defect reports are required by the FDA for any adverse reaction, side effect, injury, toxicity, or sensitivity reaction attributable to the device. The FDA Center for Devices and Radiological Health reviews premarket approval applications (PMAs) for medical devices (21 CFR [*Code of Federal Regulations*] 814.39). A Supplement to a PMA must be made if there are unanticipated adverse effects, increases in the incidence of anticipated adverse effects, or device failures necessitating a labeling, manufacturing, or device modification. There is no preemption of common law or other statutory requirements relating to device safety, including international requirements. In addition, there seems to be wide latitude among what is required, what is submitted, what is acted on, and what is approved.

We made a review of the required or mandated reports actually submitted by one manufacturer on just one product line of implantable devices. During 7 months in 2005 and 7 months in 2006, 205 adverse event reports indicated a "malfunction" or "injury" classification. The reports per month in 2005 averaged 2.8, and in 2006 they rose to 25.4. One adverse event report was evaluated in 2005 and two in 2006, so in only about 1.5% of the total cases did the manufacturer indicate that it had evaluated or determined the causes of the adverse event. Most devices had not been returned to the manufacturer after explantation, but of the 40 that were returned, 37 cases were not evaluated despite the increasing frequency of adverse events. The manufacturer did have representatives who evaluated devices at surgeons' offices, but those cases were reported as nonevaluated. Other information suggests that the devices were improved on a trial-and-error basis.

Of interest in FDA operations is the fact that manufacturers paid $232 million or 53% of the FDA's drug review budget in fiscal 2004 (Mathews 2006a). The FDA sought 66% of its drug review budget for 2007 by bargaining with manufacturers for increases in user fees (under the provisions of the Prescription Drug User Fee Act of 1992). Apparently, this gave the industry a greater role in shaping the FDA priorities, studies, and timing on such matters as proposals on new drug labels, advertising, commercials, and in monitoring postmarket safety.

Adverse Event Rates Worldwide

Hospital chart reviews in various countries indicate that adverse events in acute care hospital admissions range from 2.9% in the United States to 16.6% in Australia (G. R. Baker et al. 2004). Generally, an *adverse event* is an unintended injury or complication resulting in death, disability, or a prolonged stay in the hospital. The adverse events are those that arise during health care management. A study in Canada, covering hospital admissions during the year 2000, found an adverse event rate of 7.5% of patient admissions (G. R. Baker et al. 2004). Of these, 37% were preventable. About 5% caused permanent disability, 1.6% were associated with death, and a majority resulted in temporary disability or a prolonged hospital stay (averaging 6 additional days). It was estimated that 185,000 patients suffered adverse events annually following medical and surgical admission.

There have been numerous media stories and legal cases in Canada involving medical error. The adverse event rate in New Zealand was 12.9% of hospital admissions (Davis 2000). British studies suggest a 5% to 10% adverse event rate for hospital admissions, of which half are preventable and one-third lead to disability or death.

Wrong-Site Surgery

In a follow-up to a 1998 study of wrong-site surgery, the Joint Commission on Accreditation of Healthcare Organizations indicated that it had accumulated 150 cases by 2001. That is, about 45 cases a year were self-reported from patient complaints or from media stories. The cases included surgery on the wrong body part (76%), wrong patient (13%), and wrong surgical procedure (11%). Other organizations that have reported on wrong-site surgery include the American Academy of Orthopaedic Surgeons, which urged "controls to eliminate this system problem" (1997); the New York State Department of Health, which emphasized "enhanced communications among surgical team members" (2001); and the American College of Surgeons, which stressed the importance of teamwork and cooperative openness between surgeons and nurses (Sentinel Event Alert 2001).

Infection Control

There is an ongoing sense of urgency about infection control from an epidemiological perspective. One study applied the historical death rates for the 1918 Spanish flu outbreak to modern population data, then calculated what the current death toll would be if something similar to the 1918 Spanish flu pandemic occurred again (Associated Press 2006). The death toll or excess mortality was estimated as ranging from 51 million to 81 million, with a median of *62 million people.* Of course, the death toll would be influenced by the genetic makeup of the virus, the immunosusceptability of a world-wide population, the nutritional status in various countries, the capability of distributing a species-effective vaccine, the population density, and other factors. Estimates also indicate that good infection control practices could reduce the average rate below the 62 million projected human deaths.

Time to Rebuild

Most people in the United States believe that the health care system needs to be fundamentally changed according to an editorial in the highly respected journal *Lancet* ("Time for a Debate," 2006). The prime reasons for such criticism seem to be that a significant proportion of the population does not have access to affordable health care, that a rapid transition to newer and more complex forms of diagnosis and treatment requires basic changes, and that the source of many medical errors seems intractable.

Many reports inform that health insurance premiums are too high for many people, and a serious illness could threaten financial disaster or force the choice of nontreatment for those people. There are arguments about single-payer systems for the end-of-life-type medical care and treatment problems (Trussman 2006). Medical staff often believe that they should receive higher salaries, work fewer hours, and be subject to less bureaucracy and that working conditions should be in a less hierarchical structure. It is also privately alleged that the HMOs have had a serious repressive and regressive effect on the profession of medicine.

Regardless of the merits of such opinions and conclusions, they form a context within which medical errors occur. Therefore, they are seriously considered in formulating corrective countermeasures and preventive remedies for medical error. The changes required for an effective error control and management program should be included within the changes necessary to rebuild the current health care system if such drastic overall changes are really needed.

Our Approach

This book was prepared as a basic textbook and as a reference manual for those who may attempt to deal with and minimize medical error. It begins with an explanation of some of the important general concepts that are used throughout the book and that have a special meaning for error prevention activities. Three key areas illustrate the problems: medical services, medical devices, and analysis. But, examples are given that may have applicability to many other topics or areas of endeavor. This is followed by a description of other major methods of error prevention, including human factors, management errors, drug delivery, and communications (labels, warnings, and instructions). The appendix provides access to standards that could provide guidance or that are mandatory in terms of the compliance required by contract specifications or government regulations. A broad perspective is utilized to enhance the book's usefulness in a wide variety of situations. The concepts and techniques described are capable of isolated use on specific situations or by incorporation into a formal comprehensive program of preventive error reduction.

The error reduction and patient safety concepts presented in this book recognize the traditional and continual self-improvement efforts made in the health care industry. Nowhere else are there so many urgent short-term upgrades, updates, and transitions. There are constant subjective professional decisions in areas of uncertainty and in the adoption of new techniques from medical discovery. There is adaptation to externally imposed transformational changes. The ability to keep at the forefront of knowledge and professional practice has earned the medical community a good reputation and great respect. Based on this demonstrated capability of self-change, this book presents information that will enable self-help. It recognizes that intrusive, disruptive, costly, and wholesale reformation is neither welcome nor productive. What is suggested should be reasonable and necessary under the circumstances of diverse modern clinical practice.

This book contains a wealth of ideas, suggestions, and recommendations intended to be thought provoking and motivational. Many options are presented; some may be selected by one health care provider and soundly rejected by another. Adoption of specific countermeasures and preventive remedies will vary according to the health care system's size, budget, special needs, age and location of facilities, its mission and function, the conceptual flexibility of staff and directors, the personalities and disciplines involved, and the overall policy and rules under which it operates.

A New Meaning for Error

The term *medical error* as used in this book connotes concepts that are both broad in scope and of substantial depth. Those in the field of psychology have used the term *error* as an incorrect answer to test questions or an incorrect response in experimental procedures. This is a fairly simple use of the term *human error*. Those involved in quality inspection have used the term *error* as a cause of manufacturing process discrepancies and defects. This again is a fairly simple categorization but one that suits and serves a particular purpose. Others use *error* to describe any type of mistake, blunder, slip, lapse, or untruth. Superficiality may have some emotional value, but specificity is required for objective facts. Medical error can go far beyond simple human error concepts. The problem with earlier meanings is that customary concepts and traditional theories are limiting and bias the perspective of the investigator. The example of root cause analysis, as actually performed in some troubleshooting and problem-solving areas, suggests a somewhat simple and uncomplicated process that provides answers, but they may be of questionable utility. When used to determine medical error causation, something qualitatively different is essential. In reviewing this book, the reader may judge the merits of a broad scope of applicability, a deeper and more incisive analysis, and a process that is enriched by a variety of meaningful concepts. The slow process of reducing medical errors should be quickened if a better meaning is given to medical error, its causation, and its prevention.

Multidisciplinary Orientation

A key aspect of this book is its multidisciplinary orientation. This is necessary because of the kind of information presented and, it is hoped, communicated. Most readers will have little difficulty understanding the discrete word phrases, simplified specialty language, unique concepts, and general suggestions for the improvement of patient safety by reducing medical error. Some concepts may be more meaningful to some specialists than to others. A few concepts may elude some readers because they are foreign to their specialty and its language or operating concepts. There is no reason why an abundance of preventive measures should not be communicated to those interested. There may be reasonable objections to some ideas, but the objections may be tempered by presenting strengths and limitations, alternatives, immediate needs, and implications for further research. Anything new usually provokes both positive and negative reactions; in this case, the hope is for positive reactions and fruitful utilization.

2

General Concepts

A key question is, What causes medical error? The answer may not be as simple as anticipated. It may be complex and difficult to understand unless there is comprehension of certain basic concepts, such as those described in this chapter. What causes medical error may depend on the technique used to determine causation, the context in which causation operates, and the special disciplinary knowledge and perceptions of those involved in an inquiry. In addition to causation, there are other basic concepts that should be fully comprehended before reading and digesting the information contained in other chapters The concepts help explain the subject matter as interpreted and described in this book and the caveats on various topics found throughout the text. These include bias, standard of care, transparency, harmonization, and management of error.

Causation

An identification of the cause of an event may be required for entry in written reports of adverse events; to help select countermeasures to prevent event recurrence; for research, including gathering of meaningful epidemiological data; for data acquisition useful in determining treatment protocols; or to serve as the objective basis of technical analyses. Far too often, the cause identified is superficial, a rationalization or socially acceptable excuse, something that is not evidence based, or it is not actionable in terms of remedy. In effect, it may not be the actual or true cause.

The term *root cause* suggests that a questioning procedure has been used to determine a deeper cause, but the term has been so widely used even in casual conversations, that it has lost real meaning.

Following are descriptive categories of causation currently utilized.

General Causation

There may seem to be an immediate and apparently obvious connection between the occurrence of an error and a probable or likely cause. The perceived association between the initiating cause and the outcome event may be the result of proximity in location, time, or physical interrelationship. Such hasty conclusions regarding causation are often speculative and subject to bias and might be general in character. In terms of selecting an appropriate countermeasure, the choice or complementary match may be equally speculative, biased, and far too general in terms of cost or practicality. Specific causation is more likely to engender a remedy tailored to the situation and one that is more efficiently targeted.

The Rule-Out Process

An exclusionary process may be used to isolate the true cause. This is systematic elimination of possible causes until the most likely cause remains. It is a form of differential diagnosis that is popular. While it is a fairly simple process, there is a likelihood that it may focus on peripheral issues of causation. It may not provide definitive answers and result in only a first approximation. Since it requires informed professional judgment, it should be reserved for professional settings.

Relative Risk

A widely used criterion for causation, relative risk compares an actual risk with the background risk. Stated another way, it is the risk that includes the presence of a specific agent, condition, or factor as compared to the risk without the agent, condition, or factor. "The threshold for concluding that an agent was more likely than not the cause of an individual's disease is a relative risk of 2.0" (Federal Judiciary Center, 2000). This implies a 50% likelihood of causation, which is sufficient under federal law (i.e., the more likely than not rule of evidence).

The But-For Technique

The but-for questioning technique searches for causes. The question asks whether the adverse event would have occurred in the absence of the cause. In essence, Was the cause necessary for the result? But for the presence of the cause, the event would not have occurred. It is a logical inquiry or sequencing of pathways or circuits similar to troubleshooting an event tree. There may be a panel of questioners, each having separate occupational skills and experiences, who may use five or more rounds of inquiry (questions and answers) that successively penetrate, identify, and narrow the subject of causation.

Correlative Relationships

There may be a statistical correlation between two or more variables. The independent variable is cause, and the dependent variable is error. Correlation coefficients may be used in factor analysis to determine proportions of variance, factor loading, or independent saturations for each factor. The analysis may determine common variance, specific variance, and error variance. It is primarily used to test the ranks of number matrices if statistical correlation coefficients are available and express a relationship between the variables. Causes may be clarified using such an analysis. This method may demonstrate causal relationships (covariation) or accentuated effects, but content analysis of other information may be more productive if used on a tentative formulation basis. It is difficult to discover causation by survey methods.

Substantial Factor Test

There may be multiple causation, in which each cause, condition, or factor is assessed regarding its proportional contribution to the undesired event. It should be a substantial contribution in bringing about the error, event, or particular result. *Substantial* may be defined as something not trivial, insignificant, infinitesimal, or purely theoretical. This is a popular test because it can be used to allocate fault or cause in a proportional manner.

Concurring Cause

There may be more than one cause operating at the same time. It is a joint causation, with causes acting together or causes combining to produce an unwanted particular event. The causes are usually treated equally as long as each is a substantial cause and a reasonable producer of error. It may be used if there is a necessity of eliminating or controlling all causes that could contribute to an undesired outcome.

Intervening Causation

There may be factors or conditions that interrupt, intrude, intervene, or lie between a substantial cause and a particular result or outcome. The imposition may or may not interfere with the system under normal circumstances. It may consist of physical or psychological conditions capable of producing the unwanted event.

Remote Causation

There may be inadequate data or information to support a conclusion regarding cause and effect. The cause may be too remote to be credible. Remoteness may be temporal, by distance, or by intervening filters or barriers. In essence, remote causation is not proof of actual causation. It may be considered abnormal causation, something outside the normal environment, and considered unforeseeable in some analyses.

Proximate Cause

The term *proximate cause* has been used to signify some continuous sequence of events or causes somehow associated in time or space, unbroken by an independent or intervening cause, and without which the undesired result would not have occurred. There may be a number of proximate causes in the chain but only one responsible or immediate cause, which is the nearest or last act, event, omission, or fault. Because of its complexity, the use of this term is disfavored.

Meta-Analysis

The combining of experimental studies and quantitatively synthesizing the information is generally limited to the analysis of different research studies on the same topic. It has been used to combine nonexperimental observation data. A meta-analysis permits the identification of small dependent variable effects and may help to explain possible sources of confounding, bias, and heterogeneity. It may have use during the formulation of large-scale medical research protocols but is probably an inefficient approach to the identification of the causes of medical error as presently defined.

Risk Factors

A form of predictive causation is the fairly common use of risk factors. A physician may find that a patient has a long history of tobacco use, is overweight, has elevated blood pressure, has received a particular test result, and so on, and this combination of risk factors suggests that the patient is a good candidate for the manifestation of a particular disease entity. Risk factors may be used in screening tests, such as tests used to determine eligibility to donate blood for transfusions. The risk factors might include travel to areas where there is a high risk of exposure to the Chagas parasite, homosexual behavior that may constitute a high risk of human immunodeficiency virus (HIV), and for female donors, pregnancy that creates antibodies that can cause, in men, transfusion-related acute lung injury (TRALI). One task is to identify persons in the early stage or at high risk for a given outcome. The physician may warn a patient that cigarette smoking is a high-risk factor for lung cancer. The patient may be cautioned that chest pain (angina) may be the precursor or high-risk predictor of heart failure. The term *risk factor* is used in many different ways. Some researchers use proportional-hazards regression analysis to identify the biomarkers and their quantitative effect on overall risk.

Warning Causation

An adequate warning may cause a general practitioner to avoid prescribing a drug based on the risk information in the warning (Thomas 1992). It is effective in causing a desired action or decision. In contrast, a warning may be deemed ineffective if the physician prescribes the drug regardless of the risks detailed in the warning. It is inadequate to cause the desired result. It is also inadequate if the risks were disregarded because they were understated, minimized by sales representations, nullified by overpromotion of the drug (*Barnett v. Merck* 2006), or the causation was lost because the warning was watered down (*Tinnerholm v. Parke-Davis & Co.* 1968). Warnings must be complete, descriptive, updated, able to overcome known practices, and properly conveyed (*Martin v. Hacker* 1993). A difficult question on warning causation may be whether the pharmaceutical manufacturer can rely on a message delivered to a learned intermediary (the treating physician) or should also be conveyed directly to the patient. The key is whether the warning actually caused a prescribing physician to stop prescribing a medication. (For more information, see the Warnings and Mental Processes sections in chapter 9, Drug Delivery, pages 164 through 171 of this volume.)

Mixed Causation

In drug approvals, the question may be whether a drug causes a tumor to shrink, a disease to go into remission, a life to be extended, or the quality of life to be improved. Clinical trials provide objective data regarding causation. The drug evaluation may also consider other mixtures of causative evidence. There may be a variety of independent scientific studies published in peer-reviewed journals. There may be studies of patient-reported outcomes (PROs). There may be overseas safety reports, good practice deviations, combined studies, somewhat questionable studies, and even partially fraudulent or completely unreliable evidence. There may be backlogged postmarketing studies on critical issues. Great reliance may be placed on some evidence of causation, and other studies may be ignored. Mixed-evidence causation regarding efficacy and safety is a real-life problem for which advocacy, diverse interests, and public health objectives prevail.

There have been attempts to refine or develop new techniques for the identification of causal factors. This has included response surface methodology, statistically designed low sample size experiments, and other analytic, observational, and experimental methodologies. It has included studies of several factors, operating simultaneously and sequentially, in environments that have significant random noise, periodic irregularities, and inherent heterogeneity. Latin hypercube or other space-filling sequences have been used to cover most of the experimental regions of interest, including untested input configurations. Computer simulations or virtual experiments have been used to achieve rapid feedback on factors and response variables for robust product design, process behavior, and service problems.

Note that the selection of a particular type of causation depends on the ease or difficulty of identifying causes, the degree of precision required, and the favored methodology of the project or discipline.

Bias

There may be bias on the part of those who identify and categorize medical error and those who commit medical error. *Bias* is a prejudice, partiality, preconception, conjecture, or prejudgment that leads to misinterpretation, warped judgment, miscalculation, and hasty conclusions. If recognized, it may or may not be possible to compensate to achieve accuracy and truth.

The bias may be categorized as (1) *unknown* to the actor or others; (2) so *minor* that the actor or others believe that it is at the no-effect level; (3) *major*, which requires countermeasures, cautions, and disclosures; and (4) so *severe* that it results in adverse *intentional* acts of omission or commission.

Intrinsic Bias

A bias toward committing error may occur when those evaluating patients, drugs, or medical devices have financial interests in relevant manufacturing or marketing organizations The individual may believe that he or she is honest and can act in an impartial manner, but it takes little to initiate a bias that remains unknown to the affected person. There may be subtle psychological changes, such as personal perceptions affecting how the world is comprehended and the degree that subjective judgments may need to be confirmed or validated. Among the interests that have provided a bias is friendship, consulting and speaking opportunities for or to a company, stock ownership, and even small gifts such as free prescription pads, address labels, and key chains.

Bias may be considered a conflict of interest that could become manifest in many different ways. Should a research treatment be evaluated against a placebo as opposed to another existing alternative treatment? Compared to a placebo, there may be a significant benefit, but compared to an existing treatment there may be little or no benefit. Has the dosage and timing been selected to avoid side effects rather than optimum efficacy? Masking or hiding side effects reflects a bias, despite rationalizations to the contrary.

Extrinsic Bias

There may be external forces that create a bias. A research director may be told by management or the funding agency to conserve money and shorten the time for critical research or testing. In response, the choice may be to rely almost entirely on past research studies similar in certain given respects. The drug products may seem to differ only slightly. They are used in reliance on established past custom, practice, and proven utility. The comparisons may be used to justify approvals, early production, and competitive marketing. The extrinsic bias may serve to exaggerate benefits and downplay problems, a bias that serves an objective that could prove harmful. There may be subsequent spin about the study that creates misleading expectations. There may be reasonably anticipated error created by the extrinsic forces and resultant expressed bias.

An example from the airline industry is illustrative of how extrinsic bias can be created. One airline has a manual for flight attendants that clearly states "do not accept responsibility for an accident," referring to passenger incidents in the interior or cabin of an aircraft. Thus, in preparing an accident report or discussing an incident with a passenger, a flight attendant has to be careful when dealing with the facts, questions, and opinions. The bias may be to omit or obscure, attempt to shift blame and causation, and avoid saying or writing anything that might connote company responsibility. Insurance companies have similar cautions, intended as a liability prevention measure, to reduce uninformed utterances at the time of an accident. The difference is that the flight attendant's job depends on the exercise of bias.

Intentional Bias

There may be a conscious, purposeful, express, and deliberate effort to bias, slant, or falsify information. This intentional bias may be related to unnecessary surgery, inserting fabricated data into a research result, or simply telling a patient that further treatment and examinations are necessary when good prudent statements could reduce health care costs to the patient. Intentional bias and error are illustrated in the case studies that follow.

Knee

An orthopedic surgeon was requested by a legal defense firm to perform an independent medical examination (IME) on a person previously diagnosed with a complete anterior cruciate ligament (ACL) tear. The surgeon had previously received a number of such requests from the law firm and had generally responded in a way that minimized possible injuries, emphasized the recovery or recovery potential, and usually indicated that there would be no effect on occupational, recreational, or daily life activities. The firm was pleased by his prior negative reports, which were used in claims adjustment and litigation, and the firm continued to "send business" his way.

The person to be diagnosed had been struck by an automobile while he was in a pedestrian crosswalk. His body was rotated, a foot remained in contact with the roadway surface, and his knee was twisted. The orthopedic surgeon (IME) performed certain routine tests in the presence of the injured person's attorney. When it came to the Lachman test, which requires a force to be applied to the proximal tibia while the patient is in the prone position with the femur and tibia in an aligned horizontal position, the surgeon conducted the test with the patient sitting and the knee bent. More specifically, the Lachman tests are to be conducted with the patient in a supine position (on his back) and the knee in a 10°, 15°, or 30° flexion, not a 90° flexion (Dutton 2004, pp. 773, 774; McRae 2006, p. 156; Bickley and Szilagyi 2007, p. 551).

The ACL runs from the back side of the femur (thigh bone) to the front of the tibia (shin bone) at the tibial spine. It runs over the middle of the knee diagonally and serves to prevent the tibia from sliding out in front of the femur and provides some rotational stability. In the prone position, the tibia will feel unrestrained as it moves forward (in an upward or anterior movement). In the sitting position, the ACL is stretched over the knee, and without relaxation, the compensatory muscles will limit the degree of motion. Thus, the improper Lachman test served to mask or cover the laxity or movement indicating a tear.

The anterior drawer test was also conducted with the patient in a sitting position. The patient indicated that no other doctor conducted the knee examination in that manner. The IME report found no injury compared with the uninjured knee. The treating doctor, without the bias, found a tear and urged that a hamstring tendon autograph be used for ACL reconstruction. After repeated examinations, the patient's long-term prognosis was negative in terms of normal knee stability.

Head

A neuropsychologist selected a battery of tests to be conducted on an ostensibly brain-injured patient. The injury was traumatically induced, and earlier staged examinations revealed several areas where there were brain lesions. The task of the neuropsychologist was to challenge the degree of impairment and presence of lesions. This objective was his conclusion after scoring and interpreting a battery of tests. While discussing the results with his sponsor, several scoring problems became manifest. For each, there was a vigorous attempt to obscure, double-talk, or misstate the written scoring and how it could be interpreted. The neuropsychologist was attempting to please the sponsor and would not admit mistakes that confounded, confused, or challenged his own results. He intentionally strove to meet an objective while protecting his professional image. The intentional bias was rather obvious despite the explanations, excuses, whitewash, and attempted justification.

Performance Criteria

Physicians

The traditional or customary criterion of acceptable or unacceptable medical professional practice was derived from legal negligence (malpractice) concepts. The standard of care was *ordinary* prudence. A physician or surgeon, rendering professional services, had a duty to have the learning and skill ordinarily possessed by reputable physicians practicing in the same or similar *locality*, a duty to use the care and skill ordinarily exercised under similar circumstances, and a duty to use reasonable diligence and best judgment in the exercise of the skill and in the application of the learning. In addition, the physician had a duty to warn a patient if there was a danger to others from the patient's physical or emotional condition (including the effects of medications). When acting as a psychotherapist, there was a duty to warn identifiable third parties and law enforcement agencies of threatened violent behavior by a patient against potential victims. There was a duty to refer patients to a specialist if a reasonably careful and skilled physician would do so. There had to be the consent to treatment, and there was a duty to disclose information so the patient could make an informed decision (informed consent).

When national standards for trauma care were established, there was improved clinical care, intensive care length of stay decreased, research increased, and institutional morale increased (Erlich et al. 2002).

Note that there may be no legal liability if there is an error of judgment, a lack of success in treatment, or selection of an alternate but recognized form of treatment. The laws on professional practice vary considerably among jurisdictions and are always changing to some degree to minimize lawsuits and control damage awards. It is the essence of what is considered professional malpractice, relative to medical errors, that is important in this book.

Nurses

A nurse owes a patient certain duties of care, including having the knowledge and skill ordinarily possessed by nurses in that locality and in the reasonable exercise of care and skill similar to that ordinarily used by trained and skilled members of their profession. A nurse acting under the supervision of a surgeon may be considered a temporary servant or agent, so any negligence of the assisting nurse becomes the negligence of the surgeon.

Hospitals

The hospital has a duty to use reasonable care in selecting and reviewing competent medical staff, providing reasonable care and attention to the patient, and meeting the needs of the patient. The standard for that reasonable care is that ordinarily provided by hospitals in that locality. A hospital might procure services, collect fees, and otherwise accommodate physicians and nurses without assuming control over those services or making the physician or nurse an agent of the hospital. The hospital is legally responsible for the acts of its employees over which it exercises work control.

Nonmedical Professionals

The duties of a nonmedical professional rendering services include having the learning and skills ordinarily possessed, using the care and skill ordinarily exercised, and performing with the reasonable diligence and judgment used by other members of the profession in that locality. A technician may be an employee of the hospital, in which case the legal responsibility is vicariously transferred to the hospital.

The Pursuit of Excellence

In stark contrast with the legal standards of ordinary care in tune with general local practice, there are medical clinics, hospitals, teaching institutions, and research facilities that strive for something much higher. In the pursuit of excellence, they select and attempt to motivate the best leading-edge professionals, who are urged to strive toward an unapproachably high standard of performance. These facilities want a high national ranking and an outstanding reputation. The goal is to be the best, an example to others, a source of high-level mentoring, and a demonstration of commendable, exemplary, error-free practice. Patient satisfaction is in the forefront.

Financial Viability

A for-profit hospital may engage in competitive practices, using low-cost personnel and high-yield practices to achieve financial viability and profit for their shareholders or parent organizations. The performance standard is something just above that minimally required, however spun or rationalized. The staff may have substantial outside activities and apparent conflicts of interest. The immediate problem may be patient overload, poor or improper management and administration, or external imposition of standard procedures that are below the norm. Such institutions may be well known for their probationary status, morbidity and mortality rates, or continued medical error problems.

Medical Devices and Equipment

The performance criterion for medical devices and equipment is an absence of actionable defects, malfunctions, failures, and warranty violations. The defect may be in the design, manufacture, or failure to warn of a hazard (unacceptable risk) during the foreseeable (predictable) use. These are technical questions, the actions are generally in strict liability, and the ultimate performance criterion may be whether there is a violation of reasonable consumer expectations.

Note that for all of the above definitions, it would be wise to consult a legal specialist in the locality where a problem, issue, or conflict arose.

Transparency

Generally, all aspects of an enterprise should be transparent to the degree that everyone can see and understand what is occurring and has occurred. Medical error has often been treated as a confidential, proprietary, limited interest, or secret matter to all but a chosen few. Absence of disclosure keeps others unaware of a possible problem. This has reduced the incentives for discovery and correction of such medical errors. Well-publicized corporate fraud and corruption allegations have resulted in lawsuits, actions by the U.S. Securities and Exchange Commission, and passage of the Sarbanes-Oxley Act of 2002 (SOX 2003). There were statements that corrupt financial and accounting practices were hidden from top management, the government oversight agencies, investors, and the general public. The universal remedy advocated was transparency to enable informed and accountable management, alerted oversight groups, and timely remediation efforts.

There have been many attempts to keep medical errors a secret on the basis that those who commit an error would be more likely to reveal its occurrence if bad public relations or penalties did not result from disclosure. It was asserted that more revelations meant more opportunities to correct situations in which future errors could occur. There were rules, practices, and even statutes that accommodated maintenance of relative confidentiality of medical errors. This was to the detriment of others concerned with the reduction of medical error, as suggested by its continued high incidence, relative recalcitrance, and apparent resistance to the remedies generally imposed. A more open, informed, definitive, and systematic approach to such problems seems amply justified. Transparency is the first step in a preventive approach to medical error, but it remains a controversial matter. Transparency has been described as a primary corporate obligation (Myerburg et al. 2006, p. 2311).

Harmonization

There is an abundance of available standards that are relevant to almost every medical service, laboratory procedure, medical device, and other equipment found in medical settings. The number of standards is rapidly growing, as is their depth, detail, and coverage. Many of these standards or recommended practices have been formulated and promulgated in the European Union and by international standards bodies (see the appendix on page 205). The trend is toward adopting, specifying, requiring, or using, in part or in whole, such standards from foreign jurisdictions.

There are several problems associated with the plethora of available standards. They can become burdensome. It is a costly task to identify all applicable standards and ensure that they are implemented. Many of the standards have duplicative requirements that are time consuming and costly. Some standards are peripheral at best. A standards coordinator may be needed even in small companies or service providers. We have observed companies that ostensibly comply with one or more standards, but the compliance is sporadic, cosmetic, and illusory.

If there are too many directly applicable standards, then an attempt should be made to reduce the number utilized, consolidate in some manner those that meet company or enterprise objectives, determine which are for guidance or advisory purposes only, and determine which are key standards for which a priority exists and full compliance is necessary. There may be contractual requirements for particular standards, a marketing advantage for publicized compliance with some standards, regulations that advise the use of a set of standards, or standards that are incorporated by reference in primary standards.

There are some standards that have been *harmonized* (adopted by many countries), and such harmonized standards should be given priority since compliance has been elevated to an identified custom and practice of the industry, profession, or service. That is, a harmonized standard has become a description of minimum requirements that can be relied on by others in world trade matters as well as in domestic affairs. All standards, however, provide a guidance function and serve to retain experience that may have been gained by prior difficulties or failures. It should be cautioned that all standards reflect the compromise position of the individuals, firm, and agencies that jointly propose, argue, pay for, and vote on the provisions of a proposed standard and its amendments, supplements, and revisions.

Teamwork

It has been argued that medical specialists should operate in well-coordinated teams to enhance patient safety by reducing errors. In addition, good teamwork enhances productivity, cooperative behavior, and job satisfaction. Improved teamwork could result from special training on teamwork-related knowledge, skills, attitudes, and team management to achieve shared goals (Shojana et al. 2001; Baker et al. 2005). Since medical specialists, in particular, tend to operate in a team-like fashion, the question arises, What could be added and in what manner? The use of simulators for learning, practice, and refresher training might help in role-playing exercises concerning emergency, crisis, or rare event scenarios. It might help to change bad habits and improve cross training and might result in certified teams. The current discussions of teamwork, to reduce error potential, are still in the early stages compared to the teamwork training in military and space endeavors.

Rationalization

Medical error may infer, connote, or signify personal fault. Many individuals will immediately provide plausible excuses and reasonable explanations to justify what may be considered peer or socially disapproved behavior. There may or may not be an awareness of the mental process or motivation to substitute acceptable or good reasons for bad or real reasons. *Rationalization* is a frequent cognitive invention, cover story, contrived justification, or superficial spin for medical error to minimize the unpleasant consequences to the individual. The search for the actual cause of medical error requires probing and penetrating the aura, veils, and perceptions created by the relevant mental defense mechanisms.

Assurance Techniques

There are occupational specialists who perform functions as quasi-independent observers, inspectors, auditors, or system evaluators. They may operate from a staff or administrative office rather than having a direct line function within an organization. In other words, they assess the work of others to "assure" that it is correct or as needed. Similarly, medical error specialists cross organization boundaries, assess the error-free performance of others, and perform an assurance function. Thus, those in medical error prevention roles may look to, adopt, adapt, or modify proven techniques from assurance functions such as quality assurance or system safety. The medical error work should be compatible, to the extent possible, with the functions, employees, programs, and objectives of the enterprise.

Management of Error

Medical errors may be investigated as they occur and corrective action undertaken. This postincident response may be just one more assigned duty to a person relatively untrained, unfocused, and possibly biased regarding the causes and remedies for medical error. In other words, it assumes that almost anyone can be selected and can perform adequately in such a role. In contrast, the term *management of error* refers to an organized program specifically dedicated to the prevention of error and staffed by trained and experienced persons. A person assigned medical error responsibilities, who could be held accountable for the results, should be capable of performing informal or formal investigations and fact gathering; logically reconstructing the evidence for useful purposes; negotiating beneficial changes in sometimes-bureaucratic systems of people, procedures, and equipment; and maintaining a focus on the objective of forestalling future repeat episodes of medical error. The person should be able to develop a program that predicts error, its consequences, and remedies. There should be a quick response capability. Development of a process of accumulating relevant experience, knowledge, skills, and data is desirable. A publicized program should attract self-reporting of error in a timely positive fashion, without recriminations, and be oriented toward the discovery of beneficial countermeasures. The program should be proactive and should vigorously ferret out unreported, hidden, or cloaked medical error whether or not there was a harmful effect. The management system is to control error, not people or processes.

Caveats

Causation

Causation should be derived from techniques selected as sufficient and appropriate to isolate the real in-depth cause of error. The descriptive words and concepts that evolve during the search for causation are of value only if they permit and orient the formation of narrowly targeted countermeasures. The remedies should be capable of validation and be as effective as needed. There may be just one nexus or connection between error and a causal factor or proportional links to multiple factors.

Bias

When not discovered, identified, illuminated, and controlled, bias may cause error and pervert the search for effective countermeasures. Compensated bias does not result in a fully independent appraisal of the facts. Conflicts of interest are a prime cause of bias that alters perceptions of situations.

Performance

Performance criteria include more than the traditional standard-of-care concepts of professional malpractice that utilized ordinary care and reasonable treatment appropriate to a particular locality. The expected professional practice may be categorized by a goal of outstanding excellence or just minimal conformance. For medical devices and equipment, the criterion may be the absence of defects, malfunctions, or failures in meeting user or consumer expectations.

Transparency

Transparency is important in determining the incidence of medical error, its true causation, and the effectiveness of remedies. It is highly desirable. Nondisclosure breeds ineffective countermeasures because of a lack of awareness, incentives, specialized knowledge, and resources.

Human Factors

Medical error almost always involves human behavior. There is human interaction with medical devices, medical equipment, good clinical procedures, and health care organizational operations and decision making. Thus, reduction and control of medical errors necessarily require some understanding of human behavior under the circumstances in which human error could predictably occur.

This book contains sufficient detail about error causation and error prevention that the reader can engage in meaningful self-help activities within a health care organization. This book also provides the context, language, and guidance necessary to enable specialists in human factors to appropriately understand, tailor research, and effectively apply their skills to complex medical error problems. Such specialists may be from various disciplines, such as behavioral science, industrial engineering, macroergonomics, and human factors engineering.

3

Medical Services

Theoretical Assumptions

There are some simplified assumptions, based on modern behavioral theories and a systems approach, regarding how medical errors can be reduced or controlled in organizations such as hospitals. These generalizations have been applied with some success in certain circumstances. However, it is not possible at this time to gauge accurately what effect may be due to good intentions, good application, or the goodwill that may flow from any focused attention on a particular activity. Refinement of the process and development of skills may yield an important segment of an error management program.

Equal Status

Some hospitals have instituted an equal status program in which all personnel are ostensibly considered equal and a vital part of a team. They all have a common goal to achieve cooperatively. This contrasts with the conventional military form of command and control or a strictly authoritative caste system of human interaction or role playing. Equal status may be a form of extended teamwork derived from good surgical or research teams. Participants may take justifiable pride in teamwork endeavors and achievements rather than existence as just a quickly replaceable worker easily blamed for mistakes committed by anyone. Generally, workers rise to the level of performance needed if there is some recognition of their basic capabilities and some opportunity to excel. Without such opportunity and some semblance of equal status, the behavior of some workers may become grudging, mocking, error prone, and even marked by disdainful comments such as that doctors bury their mistakes.

Of course, equal status results in more than job satisfaction and pride in tasks well performed; it encourages subordinates to help avoid errors in the pursuit of excellence. Equal status is a relative term since there must be some chain of command in any organization, but it need not be oppressive, offensive, or demeaning or lead to insubordination. The health care professional should be alert to possible errors and have the belief that there is sufficient equal status that the possible error can be disclosed to others able to take appropriate action. The *assumption* is that some form of equal status will reduce medical errors.

Civility

Closely allied with the concept of equal status is the assumption that civility should prevail under normal, stressful, and even extraordinary circumstances. There should be a harmonious social interaction to achieve common goals in an error-free manner. This suggests a friendly, calm, pleasant, and facilitating behavior with logical decision making. In contrast, it is assumed that medical errors will result from arrogance, insulting demeanor, unethical behavior, and uncontrolled personality traits, disorders, or extreme conduct outbursts. We have been astonished what top-drawer medical specialists have been able to achieve by hands-on management with the kind of civility that denotes respect for others. The *assumption* is that lack of civility engenders or cloaks medical error.

Innocent Errors

Much of the basic human learning process is trial and error. In medicine, exploration of the frontiers of clinical practice, scientific research, and drug applications is error prone. Errors may be a predictable, foreseeable, and innocent manifestation of a developing art and science applied to a widely diverse population manifesting myriad disease entities. In some countries, there is a specific legal developmental exclusion for new medical devices because of the uncertainties and new experiences involved with innovative products. The central question is, How much caution should be exercised in the face of significant uncertainty? The *assumption* is that blameless errors may occur, but they are often correctable.

As an example, consider the recent past in our understanding of tumorigenicity and cancer therapeutics. On what basis were treatment choices made? Our early understanding of cancer was that of enigmatic uncontrollable cell proliferation with inexplicable treatment options. When a virus was associated with some forms of carcinoma, there was a search for a specific vaccine that might immunize the general population, an exposed or at-risk population, or those manifesting some early signs of malignancy (Carbone et al. 1998; Rudolf et al. 1998). There were multistage models for cell transformations related to asbestos exposure and cigarette smoking. Small-cell lung cancers were somewhat sensitive to chemotherapy and radiotherapy (Atkinson 1998). More recently, there were genes that could switch cell growth on and off (the oncogene mutations, including the *ras* genes) and tumor suppressor gene functions (the rb and p53 suppressors) with some emphasis on gene therapy (Klein 2006; Weinberg 2006). Currently, there is a cancer genome project based on the belief that the 550,000 people in the United States who died of cancer in 2004 all had cancer-causing DNA (deoxyribonucleic acid) mutations resulting from exposure to toxic substances or from damaged DNA that was inherited. The past limitations in our knowledge may have resulted in well-informed and reasonable diagnosis and treatment, but they may be innocent errors from the perspective of future knowledge.

There are those who believe that they have superior *metamemory*, the process of monitoring, assessing, and controlling memory functions. However, the very formation of memories related to instructions and knowledge serves to prescreen, influence, and modify their future perceptions. Even if they attempt to disregard certain knowledge, once the memory process has been activated, it can influence future judgments even with an intentional forgetting process (Isbell et al. 1998; Oien and Goernert 2003). In other words, preconditioning may result in innocent errors despite the actor's firm belief in independent objectivity.

Patient Involvement

There have been attempts to involve the patient more directly as an active member of the health care team. This involvement would tend to help solve informed decision requirements, ensure better compliance with treatment and care instructions, and provide the physician with information that could reduce errors. The patient should make sure that the treating physician knows about any allergies and adverse reactions to medications. The patient should be able to read the handwritten prescriptions; ask what a medication is for, how to take it and for how long; what side effects are likely to occur and what should be done about them; and what other medications and dietary supplements should be avoided.

A prescription card could be given to the patient so that the patient can cross-check the drugs given to him. When the patient receives the medicine from the pharmacy, the patient should ask whether it is the correct drug in the correct dosage. The patient should ask questions about the label, its warnings, and supplemental information. The *assumption* is that patient involvement reduces possible medical error.

An information sheet distributed by the U.S. Department of Health and Human Services in partnership with the American Hospital Association and the American Medical Association (AMA) lists the "five steps to safer healthcare" (Agency for Healthcare Research and Quality 2004a). These instructions for patient involvement were as follows: ask questions if you have doubts or concerns; keep and bring a list of all the medications you take; get the results of any test or procedure; talk to your doctor about which hospital is best for your health needs; and make sure you understand what will happen if you need surgery.

Attempts at patient involvement may be commendable, and it tends to shift some responsibility to the patient. But, most treating physicians have little time per patient under health maintenance organization (HMO) and similar payment plans. They prefer that the patient trust the doctor, understand cryptic short statements regarding diagnoses and treatment, and does not ask many time-consuming questions for which there is no financial reimbursement. Patients should comply with the five steps if they are aware of them, but assumed compliance may be illusory. It is assumed that the better educated, more intelligent, and worldwise patients will have already been involved in all aspects of their health and well-being. Those who are less able to understand may continue to view their doctor and hospital staff as omnipotent and capable of extracting knowledge on an as-needed basis. The poorer the quality of information is, the more likely there will be errors of interpretation on the part of the medical specialist.

In fact, patient involvement in their own health care has been rapidly increasing. This may be due to the availability of home-use self-test medical devices such as those that measure blood pressure and glucose levels. Some patients have implanted medical devices that require some basic understanding or special care, and many have an adjustable or programmable remote control. There is increasing use of media and Internet sources of medical information in an understandable form. Fitness programs and equipment are popular. Special outreach programs have been instituted at some hospitals. Thus, patient involvement is gradually replacing passive attitudes and nonquestioning of authoritative medical personalities. However, self-awareness and self-monitoring could lead to more self-care, and that could be a source of patient medical error.

Honesty

One of the most commendable human personality traits is honesty as slightly modified in tone to fit social niceties. Honesty is encouraged by codes of ethics and professional responsibility, as well as the common law, which reflects the fact that individuals must rely on each other in social interactions. It is socially disruptive to mislead or falsify, to commit fraud by dishonest actions, or to be verbally facile in a dysocial manner. There are those who are dishonest or untruthful and who repeatedly attempt to con others. This may suggest an antisocial personality disorder in which the rights and feelings of others are disregarded. The *assumption* is that dishonesty is accompanied by willful violation of rules, and such violations may be categorized as error-inducing events.

Although the medical profession is encouraged to be open and honest, disclosure of medical errors and an apology can be the most difficult things to do. When medical actions harm someone, the actor should simply say, "I'm sorry" and demonstrate some empathy and concern about the patient (P. Wagner 2005). One AMA report stated that physicians must offer professional and compassionate concern toward patients who have been harmed by medical error (AMA 2003). Patients want to know what happened and how future errors will be prevented.

Concepts of physician honesty now include recommendations to make *full disclosure* of medical errors that cause patient harm. The apology is part of a healing process if it includes acknowledgment of responsibility, providing adequate explanations, expressing regret and remorse, and possibly negoti-ating or discussing reparations if it would serve to soothe the patient (Laz-are 2006). The patient wants to hear that there is forbearance in the form that such error will not be repeated, and some good will result. All of this constitutes the form of honesty that has served to reduce malpractice claims by disgruntled patients. Such an effort will fail if there is a lack of sincerity, inferences of fraud, or patient perceptions of disingenuousness. Some states have legislation that make statements about fault inadmissible in malpractice lawsuits. Other states have held legally inadmissible statements of apologies and expressions of caring, regret, and consolation. There may be an assump-tion of the risk, by way of documented informed consent, for high-risk surgi-cal or drug treatments. Often, the patient accepts at face value the statement, "We are still trying to find out what happened to see if there is anything we need to correct." Some physicians will personally visit the injured patient after the incident to demonstrate a caring attitude. Above all, such actions should reflect basic honesty as a character trait.

Ombudsman

A person may be designated as an impartial ombudsman to receive mes-sages about potential medical error problems. All too often, clear and open channels of communication do not exist within an organization. There may be hurdles, barriers, and indifference at various levels of management and between departments. There may be defensive attempts to avoid calling attention to any unresolved problems in their function or operation. Bureau-cratic clutter, distortions, protective rationalizations, and clogged arteries of communication abound in many organizations. There may be adverse reactions about potential risks, avoidance about identifying specific person-alities, and negative reactions about troubled units. A direct and confiden-tial message delivery system may produce valuable information, with an ombudsman acting as a sounding board and having an assigned capability of independent investigation and remedy negotiation. This should not be confused with a suggestion system or a labor union complaint function.

The ombudsman should be easily contacted 24 hours a day so that the information delivered is fresh and unadulterated. The ombudsman should be capable of recognizing significant worthwhile information on which action could be taken. The *assumption* is that speedy action by an outsider can more effectively identify and institute timely and necessary corrective and preven-tive actions to forestall medical error. It is an alert system, but it is highly depen-dent on the skills and talents of the person assigned as the ombudsman.

Conflicts of Interest

There have been widespread conflicts of interest in the health care industry. Editors of medical journals often seem dismayed and angry that contributors to medical journals fail to disclose all financial interests and other ties that are relevant to the topic discussed in an article, research result, or evaluation. Drug sales representatives often attempt to create a bias for their products by gifts, lecture fees, or other monetary inducements. Conflicts may produce misleading opinions, result in biased research protocols, and create serious ethical issues. Institutional review boards generally neglect conflicts created by grant funding of independent scientific research because the grants further the interests of the institution. The *assumption* is that conflicts of interest may produce a bias that should be disclosed, and that failure to disclose a possible bias may be indicative of a propensity to conceal other facts, such as medical errors. It could be difficult for someone with a relevant conflict and possible bias to fully disclose known causes of error or even to perceive incipient error.

Recent examples of conflict of interest that might induce bias include the following: In the United Kingdom, the Medicines Control Agency found that two members of a panel evaluating serotonin uptake inhibitors were shareholders of a company that produced one of those drugs. The risk evaluated was whether the antidepressant drug was linked to violent and suicidal behavior. The evaluation panel was disbanded because of the conflict of interest between the panel members and the pharmaceutical industry. A new expert panel was appointed (Bachtold 2003). In the United States, a neuropsychopharmacy journal published a favorable review of a medical device for treating depression. It was undisclosed that the journal editor (an author) had financial ties to the manufacturer of the device, eight other authors were consultants to the manufacturer, and another author was an employee of the company. The medical journal published a correction, and the editor resigned (Armstrong 2006). The Food and Drug Administration (FDA) removed a member of an advisory committee because there was an allegation that the member was biased against the type of trial design utilized for a "new use" of an antibiotic in treating the infection sinusitis. The member also had been a paid consultant to the predecessor company of the drug manufacturer and had questioned the type of efficacy test used. The test made comparisons to old drugs but did not use a placebo. The drug studied might not be too much better than the placebo but was not significantly inferior to an old drug (a so-called noninferiority trial type of test). That is, a drug may be shown to be roughly equal to an older drug (perhaps slightly worse) but may not be as efficacious as a placebo (Mathews 2006b). The FDA has a policy relating to waivers of conflicts of interest that are disclosed and are considered minor.

System Dynamics

Error generally occurs in a dynamic situation, particularly at system interfaces. Medical error should be reconstructed as part of the proximate whole system, with checkpoints, surrounds, and vulnerabilities identified, such as a lack of functional redundancy. As a process-oriented throughput system, there are analogies that can be made about what industrial engineers do for the industrial manufacturing process. They optimize the flow, between a series of functions, to enhance productivity while reducing waste and rejects from material, machine, and human error. There can be better comprehension of potential errors if the context or system within which they occur is understood. The *assumption* is that a systems-oriented approach to medical error would reveal much more than looking at a few isolated discrete elements of the system.

Medication Errors

Extent of the Problem

Among the most frequent medical errors are those involving medications. The drug system involves procuring the drug, prescribing it, dispensing it, taking it, and monitoring its effects. Each week, four of five adults will use 1 of 10,000 prescription medicines, 1 of 300,000 over-the-counter drugs, or countless dietary supplements. One-third of the adults will take five or more different medications. Thus, medication errors could involve a large proportion of the population.

In July 2006, the Institute of Medicine (IOM) issued a report, "Preventing Medication Errors." It concluded that "when all types of errors are taken into account, a hospital patient can expect on average to be subjected to more than one medication error each day." The IOM report stated that at least 1.5 million preventable adverse drug events occur in the United States each year. Each of these errors added to the cost of a hospital stay. In addition, there were the patients' lost earnings, possible emotional reactions, pain and suffering, and other health concerns.

Prevention Strategies

The 2006 IOM report discussed the following strategies intended to reduce medication errors: First, there should be a shift from provider-centric services to a communications partnership between patients and their health care providers. That is, patients should take more responsibility for monitoring their medications. A recommendation was made to standardize and improve information leaflets provided by pharmacies, to make more drug information available over the Internet, and to develop a national 24-hour telephone helpline. Second, doctors should make better use of information technologies. This includes point-of-care reference information on the Internet or personal digital assistants, electronic prescriptions to avoid errors from handwritten orders, and use of electronic prescriptions to check automatically for drug allergies, drug–drug interactions, and excessively high doses. Third, there should be improved labeling and packaging. The information sheets provided with drugs should be redesigned. Drug nomenclature should be improved. Fourth, accreditation agencies should require more medication management training.

Individual Remedies

Not all errors make their way to the patient; some are intercepted. One early study indicated that 39% of all adverse drug events were traced to incorrect orders, but half of them were caught by pharmacists or nurses (Brennan et al. 1991). Nurses made 38% errors, but only 2% were caught because of fewer checkpoints between the nurses and patients. A frequent corrective action is to insert manual or automatic checkpoints in the drug delivery system to catch and nullify errors.

Another study recommended that drugs be individually bar coded to reduce medication errors (Wright and Katz 2005). It also suggested that pharmacists could remind the prescribing physician to lower the dose or order an alternative medication to reduce side effects. Nurses could be alerted to discover undocumented allergies or question dosage levels for a particular patient.

Nine vials of phenytoin, an anticonvulsive, were found in the heparin bin of an automatic dispensing cabinet. Both products resembled each other in size and labels (Cohen 2005). In such a situation, the pharmacy could have affixed supplemental labels emphasizing the contents. Another solution for the pharmacist would be to purchase each drug from a different manufacturer.

In pediatrics, colorful wristbands can serve to identify each child's weight category. This enables clinicians to quickly choose equipment or drug dosages appropriate to a child's weight and size ("Colorful" 2005).

Computerized Systems

The use of bar codes has enabled computers to perform many functions that could reduce medication errors. Patients could wear bar-coded bracelets that could be read and compared with the bar codes on drug containers. A match could eliminate the wrong medication, wrong dosage levels, or wrong patient. In terms of drug supply management, smart labels permit the traceability of drugs by product code, lot number, manufacturer, and expiration date. They permit automatic reordering and stocking. Also, the patient is provided with a printout of the drugs actually administered to ensure correct medications, timing, and dosage.

One drug management system involved a computerized prescriber order entry to avoid illegible handwritten orders that could lead to ambiguity, guessing, and error. The electronic order entry would accept orders in a standard format, with strict criteria regarding what is accepted; had default values; and issued reminders and warnings. The order entry system should be designed so that it cannot be used incorrectly and introduce errors into the system (Grissinger and Globus 2004).

The system also involved automated dispensing cabinets that sounded an alarm if incorrect drugs were placed on the cart. High-alert medications, including heparin, wafarin, morphine, and potassium chloride, should not be stored in such dispensing carts. A pharmacist should screen any medication before it is placed on the cart. Look-alike and sound-alike drugs should not be placed adjacent to each other or in the same drawer. If unit-dose dispensing is not used, then special care is needed to screen every medication order. Such bar code systems have substantially reduced medication errors.

Radio-Frequency Identification Device Technology

There is an increasing use of radio-frequency identification devices (RFIDs). This involves the use of a small tag placed on items or embedded under the skin of animals and humans. The tag contains a microprocessing or a microcontroller chip and antenna (a passive tag), and it may include a battery (an active tag). When in proximity to an electromagnetic reader or scanner (an activator), there can be access to substantial real-time information. For example, if a patient is brought to a hospital emergency room (ER) and cannot communicate, then the ER workers can check the patient with an RFID reader or scanner. An activated embedded tag or chip can provide a quick readout of substantial medical and personal information.

The RFID technology is becoming commonplace. An automobile may be immobilized until the vehicle's RFID reader detects the correct tag in the driver's ignition key. It has been used to identify and track animals. In the manufacturing and distribution of products, items such as components, assemblies, and products have been tagged to reveal immediate information on current status, location, prior movements, testing, inspection, storage, distribution routes, and eventually, conditions of use (Peters and Peters 2006a, pp. 15, 148). Previously undetected errors are revealed by computer analysis.

People have implanted RFID tags to provide access to security areas in buildings. Plastic smart credit cards may contain a tag to reduce theft and unauthorized use. There are multilevel security and privacy features that may be built into smart cards, shielding to prevent activation, antizapping safeguards, and tamper-proofing features. There are also tiny pellets called *taggants* that can be mixed with chemicals or medications that serve the same purpose of tags affixed to drug containers.

It is apparent, in terms of medical error, that RFID-enabled computer systems may be configured to define, inspect, detect, alarm, reject, or institute corrective action for undesired events, conditions, or actions. The technology has direct application to reducing and controlling medication errors in terms of both prevention and correction of dosage levels, type of medication, order conditions, drug interactions, delivery to the correct patient in a timely manner, cross-checking of side effects and past history, and, in general, providing cost-effective automated systems with substantially reduced error.

Note that, for objects less than 1 mm in size, there are fluorescent dye tags called molecular computational identification devices (MCIDs). Millions of distinguishable molecular tags are possible using fluorescence output patterns for identification (Freemantle 2006).

Clinical Conversations

The clinician conducts interviews or structured conversations with the patient and prepares an organized problem-oriented health history. The record may be incomplete or nuanced, contain diagnostic hypotheses, relate the patient's story, state the need for an interpreter, describe prior diagnoses and treatment, and provide patient-oriented goals (rather than just provider-oriented goals). The error-prone behaviors of some health care specialists have included the following: failure to review the medical chart or records before treatment, failure to describe in the records any informed consent conversations, ignoring reasonable requests of the patient, and failure to conform to a negotiated and mutually acceptable plan of treatment.

The health provider should wish to do no harm and act with beneficence regarding the patient's wishes. Error-prone behavior is to do what the provider deems best unilaterally and to ignore the patient's right to determine what is in his or her best interests. A patient might have been initially assessed by an empathetic listener, but a specialist to whom the patient is referred may not be as socially or detail oriented and may disregard notations in the records. The specialist and the primary care physician may duplicate drug prescriptions and offer conflicting opinions to the patient. Symptoms elicited by one may not be communicated to the other by way of updated records and reports. The medical records should be reviewed to determine medication history, particularly polypharmacy suboptimal prescribing, prescription nonadherence, confusion from drug labels and names, any pharmacodynamic changes in metabolism and elimination of drugs, and past interactions with over-the-counter or recreational substances. A pure focus on clinical thinking for an immediate objective may overlook potential error-producing events.

Comparative Risks

The frequency of medical errors may seem disconcerting, embarrassing, or frustrating to medical specialists who have great pride in their profession. They should be aware that such human errors occur in many professional and technical activities. Over 60% of all airline flights have some kind of error in cockpit activities (McCartney 2005). The International Civil Aviation Organization has urged all countries to develop threat-and-error management programs. Aircraft pilots use checklists to reduce mistakes. Errors include wrong directional heading, wrong speed, and wrong configuration of flaps. There may be distractions, lack of communication between members of the flight crew, and other error-producing situations.

Errors and resultant defects constitute more than 45% of all manufacturing problems (Price 2005). The errors may be associated with lapses of concentration during extended uneventful vigilance, operations outside the normal envelope, miscues during transient conditions, and lack of proper procedural sequencing if there are poorly defined procedures.

In 2005, a railroad freight train crashed into another train parked on a railroad spur. It pierced a chlorine-filled tanker car, the chlorine gas cloud drifted over nearby houses, and nine people were killed and dozens injured. The Federal Railroad Administration concluded that human error was to blame (Machalaba 2006). Hazardous material releases from railroad crashes have primarily occurred from tank failure, track failure, or human error. The proposed remedy was to "harden" or replace about 12,000 tank cars that haul chlorine or anhydrous ammonia (used as a crop fertilizer). Hardening means that the metal thickness on the tank cars would be increased by 25%, and there would be extra padding on the tank car ends and special padding for the valves. Trade groups suggested that there should be improvements in railroad maintenance and operations. Railroad cars routinely carry substances of far greater risk than chlorine and ammonia, and some trains pass through major cities. This has been called a national security problem. Some railroads have significantly reduced human error by design modifications to the system (Peters 1999, pp. 65–70). However, there is still some residual risk from human error.

A test kit error occurred when the influenza A strain H2N2, the cause of the 1957 Asian flu pandemic, was sent in a panel "accidentally" to thousands of test labs (Enserink 2005). All the kits had to be destroyed by the laboratories in 19 countries. The World Health Organization responded by saying that they would issue a recommendation to "bump up safety procedures" by those working with the virus.

In August 2003, there was an extensive blackout of the North American power grid that had an impact on 60 million people. There were deficiencies in dispatcher recognition of deteriorating conditions (an error of omission) and in taking proper effective action (an error of commission). As a result, utility power grid dispatchers now can train on simulators that intentionally introduce human errors and instrument failure to evoke a loss of situation awareness (Dagle 2006). The objective of introducing system errors and failures was to achieve better recognition of false data and misleading scenarios and to improve troubleshooting of off-normal conditions even when alarms fail to operate.

The Three Mile Island nuclear power plant partial meltdown of a reactor core in 1979 was caused by a combination of human error, design deficiencies, and component failures. The control room operators were trained to look for symptoms, not to understand root causes. They delayed taking action (an error of omission). Training now emphasizes recognition of bad information and action to prevent system problems. This includes operator cross-checking of instruments, design avoidance of common modes of component failure, training on hacker threats, and standardized approaches to emergency situations.

The Bhopal tragedy was caused by multiple human errors of neglect, omission, and misunderstanding. Human error has been involved in building flaws, food safety mistakes, automobile crashes, and maternal child care (Peters and Peters 2006a). Thus, medical error is simply something to be expected, and special control measures are needed to reduce what could be costly mistakes.

Infection Control

Sources of Contamination

In 2000, a Patient Fact Sheet issued by a federal agency stated, "If you are in a hospital, consider asking all health care workers who have direct contact with you whether they have washed their hands" (*Twenty Tips* 2000). Six years later, in 2006, hospital-acquired infections were estimated to occur in 10% of all acute care hospitalizations (Green-McKenzie and Caruso 2006). Another source indicated in 2006 that 5% of patients contract an infection after they are admitted to a hospital (Landro 2006a). Both sources agreed that approximately 2 million patients each year acquire an infection directly from the hands of health care workers or indirectly, such as from catheters that have been in contact with the contaminated bodily fluids of other patients.

There may be pathological microorganisms on gowns, bed rails, bed linen, furniture, and other objects in the immediate vicinity of the patient. There is a transition from multiple (ward) or double-occupancy (semiprivate) rooms to single occupancy to reduce such transfer of contamination from infectious agents. There are also changes in access to patients as described in this chapter.

To fight bacteria, there are resistant bedsheets, gloves that release disinfectants, catheters coated with antibiotics, bacteria-resistant lab coats, disposable cloths saturated with chlorohexidine for patient bathing (rather than just soap and water), quicker tests to identify bacteria, stethoscopes wiped with alcohol, intravenous changes every 3 or 4 days, and rapid reactions to localized outbreaks of bacteria.

There are two problems. First, there is a recognition of the obvious risks from infected draining wounds or wound dressings and the need for presurgical antisepsis. Other possible infection sources may not seem important. Second, it is discomforting for health care workers first to decontaminate or wash their hands, then put on sterile gloves, subsequently remove the gloves for every patient or contact with inanimate objects, and finally again clean their hands after glove removal. Many believe that there are just too many time-consuming and bothersome steps in such infection prevention efforts. It is important to note that each person, patient, or provider sheds about a million cells a day that can contain viable microorganisms. What are the tolerable limits of this source of risk?

Hand hygiene includes use of plain soap and water, chlorohexidine, chloroxylenol, hexachlorophene, triclosan, quaternary ammonium compounds, and alcohol-containing antiseptic agents. The hand-washing facilities should be convenient and easy to use; require no contact with other dispensers; have available rinses, foams, or gels that are used with sufficient contact time and be in dispensers that are regularly refilled. Agents should be applied with aggressive hand motions, and recontamination from faucet, handles, towels, or dispensers should be avoided.

Many hospitals have staff monitors who provide surveillance, observe health care workers, and report violations of hand hygiene rules. Compliance varies but averages about 50% nationwide. As smarter bacteria evolve, more aggressive use of different techniques is expected. The Joint Commission on Accreditation of Health Care Organizations requires hospitals to have an infection control program in which hand hygiene is an important element.

Given what could be done or is done in the name of infection control in some hospitals, should those remedies be considered as criteria for defining what constitutes medical error? Are all current countermeasures to medical error still too experimental or speculative regarding their real substantive value? Is it appropriate to wait for confirmed anti-error procedures and future developments? See chapter 6, pages 99, 100, and 118, on the relation of defiant actions to infection control.

Use of Antibiotics

Disinfectants, antiseptics, or germicides act against bacteria by denaturation, breakdown of cell walls, or interference with reproduction. They are effective against many bacteria, fungi, and viruses. A different approach is the use of antibiotics to fight bloodstream infection. However, the unnecessary use of antibiotics can lead to drug-resistant or smart adapted organisms. There has been increasing bacterial resistance to broad-spectrum antibiotics. How these infectious disease control agents are used may be called drug errors or reasonable and good therapeutics. Also, see the section on biofilms in chapter 4, pages 64 and 77, of this book.

An orthopedic surgeon performed a high tibial open-wedge osteotomy and the operative report and status reports indicated that no antibiotics were administered presurgery as is generally recommended. An infection occurred, and there was nonunion of the wound. A 1- or 2-week course of antibiotics was ineffective, so it was continued for about 1 year. At that late date, there was a consultation with an infectious disease specialist. A cocktail of antibiotics was prescribed as an intensive treatment of the infection. There had been no early consultation with other specialists. There had been removal of hardware because of nonunion. No magnetic resonance imaging had been taken to rule out osteomyelitis. The infection was under control and ceased at about 18 months postsurgery. At that time, there was evidence of osteomyelitis, and the history of a high level of pain continued after wound healing. In essence, the surgical infection was improperly diagnosed and treated for more than a year. This illustrates that there may be combinations of multiple medical errors that occur over a long period of time.

Medical Waste

Generators of medical waste have legal responsibility for the waste generated, its disposal, and its final destruction. Health care waste includes discarded sharps (such as objects that can cut or puncture the skin); infectious waste that is expected to contain pathogens (such as bacteria, viruses, parasites, or fungi); pathological waste (such as human tissues, fluids, and body parts); and chemical wastes (such as drugs, vaccines, cleaning chemicals, radioactive materials, and pharmaceuticals no longer required). The source of medical waste includes the major generators (such as hospitals) and minor generators (such as medical clinics, doctor's offices, laboratories, health centers, and veterinary practices). There should be an effort to minimize the generation and disposal of medical waste at its source. This includes the selection, purchase, and use of supplies that generate less medical waste. It includes use of recyclable products and materials and the segregation of high-risk materials (Obid 2006). That is, there should be special management control over medical wastes if error could have adverse results.

A medical waste management program is desirable, with constant updates and annual refresher training. Bloodborne pathogen standards require sharps (such as needles) to be disposed of in puncture-resistant, properly labeled, and specially designed sharps containers. This disposal method is used for hypodermic needles, syringes, infusion sets, lancets, suture needles, scalpel blades, and glass, box cutters, and knives contaminated with blood. Sharp disposal containers and spill cleanup kits should be located in each examination and procedure room. Vandal-resistant containers should be in point-of-generation locations such as employee restrooms, cafeterias, mailrooms, and storage facilities. Injuries can occur any place at any time with the release of infectious bloodborne material.

Outside of health care facilities, there is home self-injection of prescription medications for diabetes, hepatitis, allergies, and other diseases that have generated over 3 billion sharps each year in the United States. Have these patients been instructed in proper waste disposal methods?

Regulated biohazard medical waste includes items saturated with blood, semen, vaginal secretions, or fluids visibly contaminated with blood. Biohazard spill, cleanup, and disposal kits or body fluid spill cleanup kits should be located where needed.

In 2000, a national study of water streams found that 80% contained pharmaceuticals, hormones, and organic wastewater contaminants (Rubinstein

2007). A similar study in Canada found nine different drugs in the water near drinking water treatment plants. The drugs included ibuprofen and naproxen (pain killers) and gemfibrozil (a cholesterol-lowering medication). Newspaper accounts claimed feminizing estrogenic compounds were found in the water, and that dedicated safe disposal procedures are needed for unwanted, out-of-date, unsalable, and over-the-counter medications. There have been newspaper reports of hypodermic needles and small plastic bags of medical wastes washing up on public beaches as the ocean tides come in. Hazardous waste disposal facilities are to be used instead of landfills, incinerators, and waste-to-energy facilities.

Some hospitals had their own incineration to disinfect the waste that required transport and disposal elsewhere. But, there were mercury emissions (from thermometers, batteries, and amalgam) and dioxin (from polyvinyl chloride). Thermal or pyrolytic incineration is still the prime method of treating medical waste, but it is in centralized facilities with gas emission or flue-cleaning equipment. The waste is thermally decomposed into ash and gases. There may be a secondary combustion chamber to complete the process. Hospitals use steam thermal disinfection with autoclaves, advanced autoclaves, and retorts (with no steam jacket). The disinfected waste should be analyzed, on occasion, to determine if it can be handled as municipal solid waste. Chemical disinfection may be used for pretreating blood, urine, sewage, and infectious waste. The chemical residue may present environmental problems. Landfills for treated medical waste have special requirements to prevent pollution by using barriers and liners that also permit leachate collection.

The disposal of medical waste involves compliance with government regulatory requirements, technical options for treating various types of medical waste, and a wide variation in economic costs. At each decision and operating stage, the potential human errors should be appropriately considered and the consequences evaluated.

Failure to comply with appropriate medical waste procedures and regulations is considered a medical error because the point of generation is in a hospital, medical clinic, or physician's office, and waste control starts at the point of generation.

Staff Immunization

Mandatory immunization of all health care workers is advisable to stop the spread of infectious diseases within the closed space of a hospital. The health care workers are in close contact with patients and other staff members, so transmission of disease is clearly foreseeable. A firm policy indicating the necessity for immunization is a strong and direct message that there is a culture of infection protection that is important. A clarion explanation to justify staff immunization is the fact that influenza protection reduces patient mortality (Todd 2006). Mandatory immunization may be part of a disaster plan so health care workers should not be absent or suffering at home from a pandemic illness.

Nosocomial Pneumonia

The importance of nosocomial infections is suggested by the fact that there were more than 100,000 cases of hospital-acquired infections in England each year (Mayor 2000). This led to 5,000 deaths. In Canada, 70% of all hip fracture surgical patients did not receive appropriate antibiotics (Zoutman et al. 1999). In the United States, approximately 10% of all hospital patients acquire a clinically significant nosocomial infection, and 20,000 deaths result from the infection.

The hospital-acquired pneumonia incidence in the United States is about 2 to 6 per 1,000 hospital admissions, with a 20% to 30% mortality rate. One frequently ascribed cause is the use of mechanical ventilators without elevating the head of the bed. But, respiratory pathogens abound from poor oral care since dental plaque is a good growing medium for microorganisms. It is generally recommended that the patient's teeth, oral mucosa, and tongue be brushed every 4 hours with a moistened suction toothbrush. Dentures should be brushed daily and cleaned weekly. The reduction in pathogens has been found to improve swallowing and coughing reflexes (Weitzel et al. 2006). Despite staff shortages, failure to reduce respiratory pathogens should be considered a medical error.

Nosocomial infections are the result of the chain of transmission from patient to patient, a high prevalence of the pathogens, and the presence of a compromised host. The hosts may be susceptible to infection because they are immunodepressed, have broken skin, or display wounds. The transmission may result from a disregard of protocols intended to inhibit or prevent contact (to break the chain) between infected patients and hospital workers, visitors, and noninfected patients.

The pathogens commonly include *Staphylococcus aureus* (a gram-positive, spherical-shaped bacteria that may be methicillin resistant); *Enterococcus* species (a gram-positive bacteria that may be vancomycin resistant); *Escherichia coli* (*E. coli*; a gram-negative bacteria); *Pseudomonas* species (a gram-negative bacteria). The site of nosocomial infections may be the urinary tract, surgical wounds, the respiratory tract, the skin, the blood, the gastrointestinal tract, or the central nervous system.

Prevention in general includes avoiding direct contact with patients and their body fluids, such as blood, semen, vaginal tissue, cerebrospinal fluid, and other fluids. It includes wearing barrier gloves, avoiding sharp punctures, observing aseptic techniques, washing hands frequently, isolating patients, filtering air, using single-use disposable items, and actively overseeing an aggressive infection control program.

Nosocomial infections remain a serious problem despite the infection control programs and the human body's natural defenses (such as white blood cells and fever), barriers (such as skin), and specific mechanisms (such as antibodies). The opportunities for human error abound in the causes and preventive measures used to control nosocomial infections.

Call Lights

The patient's use of call lights may seem bothersome at times. The response time for call lights and alarms may vary according to the availability of registered nurses, licensed practical nurses, certified nursing assistants, patient care technicians, and patient service partners. Caregiver burnout, job dissatisfaction, fatigue, overload, preoccupation, and general inattention may increase the response time to a lifeline connection. The call lights may be considered something to do with food service, housekeeping, cleanliness, the need to use a toilet, or some form of meaningless attention getting. The patient should have a responsive means of immediately obtaining help as part of pain control, general well-being, the peace of mind from a belief in the availability of help when needed, and satisfaction with the quality of care given. The nurse's rounds should be supplemented with call lights, alarms, and the capacity for continued observation from a nurse's station. Periods of complete isolation should be avoided.

Since many patients stop taking lifesaving drugs shortly after hospital discharge, responses to call lights could include repeated discussion of the importance of taking the prescription drugs at home, even if the patient has a feeling of well-being. During the discussion, the patient's current condition might be better evaluated than by other means if concern is manifested and rapport achieved. The explanation of continued drug use may indicate that more than 10% of heart attack victims discontinue their drugs (such as aspirin, beta-blockers, and statins) after about 1 month and become a higher risk for clogged arteries and bypass surgery. Similarly, diabetes patents fail to take their medications (such as hypoglycemics, blood pressure drugs, and statins) and become a higher risk for rehospitalization. Just one discussion by the treating physician seems to be insufficient to prevent some patients from stopping their medications when they feel good. (For more information on call lights, see the managerial section in chapter 7, page 121.)

Drug Resistance

There is a need to identify patients who are resistant to certain drugs. Some patients have a high risk for colonization with methicillin-resistant *Staphylococcus aureus* (MRSA) or vancomycin-resistant enterococci. Such risks may be determined from the records of previous hospitalization within 1 year or, better yet, the patient's memories (Kayyali 2006). This emphasizes the need for better public health tracking, past and future, with monitoring protocols as enabled by modern computer systems.

There is considerable discussion of drug-resistant bacteria and viruses in this book. The resistance may be fairly common, extensive, multidrug, or to a particular strain. The important considerations are the use of other antiviral or antibacterial drugs, the judicious use of vaccines, effective isolation, and quick quarantine. The belief may be that global mobility helps to spread infections; that is, the Boeing 747 aircraft is an important disease vector. Such vectors may be of limited importance in controlling pandemics. There is also a strong belief that when patients do not finish a full course of medication or receive drugs that have limited potency, a gene mutation or drug resistance develops.

Drug-resistant forms of tuberculosis are a serious challenge in some countries (Aziz et al. 2006). More information is needed to develop appropriate control measures. Drug susceptibility tests have been conducted on four tuberculosis drugs (isonizid, rifampicin, ethambutol, and streptomycin) in this recent investigation. Such studies indicate the need for surveillance of drug resistance since such information is a vital component in monitoring infectious diseases.

Touch Surfaces

The antimicrobial properties of *copper* were known many years ago by the Egyptians, Greeks, Romans, and even the Aztecs. Uncoated copper and copper alloys can inactivate *E. coli* (O157:H7), streptococcus, and MRSA. In 2006, the process of registration of such health claims started under the Federal Insecticide, Fungicide, and Rodenticide Act (Michaels 2006). Copper alloys used for human touch surfaces, such as doorknobs, prevent continued cross-contamination. In health facilities, the surfaces needing continued antimicrobial protection include bed rails, furniture hardware, medical monitoring equipment, intravenous unit stands, sinks and faucets, and work surfaces. Disinfectants and antimicrobial coatings have limited duration or efficacy. Copper also has an antiviral (viricidal) effect on Influenza A (H1N1) and may have a wide-spectrum effect. It has an antifungal effect on *Aspergillus niger*. Copper alloy touch surfaces in nursing homes, locker rooms, gymnasiums, schools, and transportation vehicles could reduce community-acquired infections.

The many copper alloys include brasses, bronzes, and copper-nickel alloys. When their antimicrobial functions are desired, surface plating or lacquering should not be used. Copper alloys generally have a pleasing color and have favorable characteristics such as malleability, formability, strength, weldability, and nonmagnetic properties. These attributes, such as mechanical strength, may be enhanced by the metals selected for the alloy. There has been considerable experience with copper toxicity. For example, a long-time use is for copper sulfate pentahydrate, which has been used as an agricultural fungicide, algicide, and bactericide, and in insecticide mixtures.

However, further research is necessary to determine the optimum alloys for health care facilities, their surface finishes, their corrosion resistance, and comparative efficacy in terms of antimicrobial intensity and duration for given pathological agents.

There have been studies of the mechanism of action of copper against pathogenic organisms. Normally, the human immune system operates by sequestering bacteria in macrophages, which contain appreciable amounts of copper that control the bacteria. Bacteria require some copper to survive, but too much copper creates reactive oxygen species that can be harmful. In about 200 bacterial species, including many antibiotic-resistant bacteria, there is a copper-sensitive protein that binds on the DNA and controls copper by pumping out high levels of copper to achieve homeostasis. Such studies may lead to novel genes that control and achieve high levels of bacterial copper. The research effort is to identify the genes responsible for copper regulation and their role in copper vulnerability, biofilm formation, and bacterial virulence (Everts 2006b).

Silver has been known to retard the growth of bacteria and algae ever since the days of the ancient Romans and Greeks. It has been called a broad-spectrum antimicrobial agent (Hermans 2006). There are more than 10 silver-containing wound dressings, although evidence of their efficacy is conflicting in terms of concentration, amount, compounding, toxicity, adverse effects, and chemically available silver. However, there are reports that silver is effective against MRSA and vancomycin-resistant enterococci. Silver ions are apparently absorbed into the wound site; they bind to bacterial cell membranes and are transported into the cell, where they bind to DNA, impairing cell replication, and bind to intracellular enzymes, inactivating them (Hermans 2006). Testing of wound care compounds and procedures is desirable, and there is a grading system that could be used (http://www.cebm.net/levels_of_evidence.asp). The risk of complications should be avoided when using silver-containing dressings (van Rijswijk 2006).

There are silver nanoparticle products that have inhibited the growth of *Aspergillus niger* molds. They have diameters of about 80 nm with high surface (reactive) areas. In wallboards, there may be nanotube-encapsulated nanosize biocides that have controlled release either slowly or when moisture is present. They have been proven effective against *Aspergillus*, *Penicillium*, and *Stachybotrys chartarum* (black) molds, and wood-decay fungi (white and brown rot). There may be an on-demand release lasting several years.

Recent research with *polymer paints* has shown that they can kill infectious agents on surfaces such as door handles (Arnaud 2006b). A polymer, such as polyethylenimine, is given an electrical charge, and that causes the polymer to stand upright on a surface. The upright polymer spike then interacts with the lipid bilayer envelope of a virus by poking holes in the envelope and inactivating the organism. The spike is formed by a hydrophobic molecular chain. Tests have shown that such a coating can inactivate both pathogenic bacteria and several strains of the influenza virus. The killing process takes about 5 minutes, and the current research is getting the process to work faster while preventing any evolving resistance by the microorganisms. Further near-term research is necessary to develop a coating that is stable and has resistance to cracking. Research has illustrated that there is considerable effort in some areas to locate surface antimicrobial agents.

A so-called self-cleaning paint or coating has been used in Europe. It has a nonstick surface to repel water, dirt, and other contaminants. Small spherical nanoparticles of titanium dioxide have enough oxidation power to kill bacteria (Yoders 2006). There is ongoing research in the United States to develop low-maintenance, bacteria-resistant paints using combinations of silicones, fluorocarbons, calcium carbonate, and titanium oxide.

There is a need for effective antimicrobial agents for street clothing used in hospital settings, scrubs, drapes, gowns, masks, bedding, gloves, seats, instruments, equipment, mattresses, bandages, air conditioners, heating and ventilation filters, upholstery, fabric-covered wallboard, laundry machines, and other objects that might become contaminated and serve as a source of infection. Such infection sources may be located in hospitals, clinics, nursing homes, dental offices, or equipment biofilms. The antimicrobial coatings for fabrics have used cyanide- and arsenic-based biocides. However, these are toxic; they slowly migrate to the surface, can be washed away, are worn off, and become gradually depleted and ineffective.

The use of *nanotechnology* has permitted the permanent embedding of nanoparticles into the surface of a product or object. If the recommended cleaning procedures are used, then the nanoparticles are not washed away, worn off, or depleted over time. Testing of some of these products indicated that 99.9% of resistant *Staphylococcus* bacteria are killed within 30 minutes ("Reduce" 2006). The nanocomponents used are about 10 nanometers long. A nanometer is 1 billionth of a meter, and for comparison purposes, a human red blood cell is about 2,000 nanometers long. The biocide chemicals vary in molecular formulas, so test results are important in determining their effectiveness and kill timing when used as antibacterial, antiviral, and antifungal agents. The benchmark question remains: Are disposables more cost-effective?

Note that size is relative. Biocide nanocomponents are about 10 nanometers long. The human neural synapse gap may be only 10 to 50 nanometers in separation distance, but it contains more than a hundred postsynaptic proteins that serve as gatekeepers, neurotransmitters, and connections to signaling proteins inside the cell. The biocide tubules can be tightly packed within a matrix, thus providing a dense, long-lasting effect.

Whose error is it that there is so little hard evidence regarding *wound dressings* that could support good clinical practice rather than trial and error? Some information is available at www.woundsresearch.com/wnd, but preventive monitoring should occur following wound cleaning, removal of necrotic tissue, deep swab culture, or needle aspiration; laboratory findings help decide whether systemic or topical antimicrobial therapy is necessary.

In the future, it is expected that a wide variety of antimicrobial products will reach the marketplace. Even the consumer market has antibacterial toothpastes, household sprays, and hand lotions. Some medical devices have antibacterial silverplated components. Wall coverings have been treated with antifungal agents. Any coating is a surface modification with claimed intrinsic effects and other surface effects. There are special lighting systems and filtered air conditioning systems. It is important to review independent test laboratory reports indicating the actual effect of such products on specific bacteria, viruses, and fungi.

Lessons Learned

There is a well-known history and public fear about infectious diseases. This resulted from the Black Death pandemic of 1346–50, the Spanish "flu" in 1918–19, the 1957 serial transmission influenza A virus pandemic, the fungal infections of the Irish potato crop in the 1840s, the foot-and-mouth infections of livestock, the current threat of a bird flu epidemic, and the annual influenza injection publicity that still occurs. There are public expectations in terms of preventive medicine, public health, and the control of infections.

The fear of highly infectious diseases is reflected in what occurs when there is a human outbreak from the Ebola virus or hemorrhagic fever transmitted among and by the gorillas in Africa. An epizootic wave of Ebola in national parks and wildlife sanctuary disease reservoirs spread predictably and killed 5,000 gorillas in 2006 in Gabon and the Republic of the Congo (Bermejo et al. 2006; Vogel 2006). There were human outbreaks in 2001 and 2002, with lag times to death of 12 days. The fear of a human epidemic resulted from media stories and efforts to contain the spread of Ebola species. There were even discussions of a vaccination campaign for wild apes.

Controlled sources of infection can weaken, and a suppressed infectious disease can reappear. The control of malaria by *mosquito suppression* may have cycles of effectiveness and ineffectiveness.

Malaria is caused by a parasite, the four species of the organism *Plasmodium*, spread by the bite of the infected female *Anopheles* mosquito. Mosquito suppression involves the use of long-lasting insecticide sprays, mosquito netting, and mosquito repellents. Preventive drugs include chloroquine, doxycycline, and mefloquine. Some mosquito strains have developed drug resistance. Malaria remains a common infection in the tropics, so suppression is only partially effective. Transmission also occurs from blood transfusions that are contaminated and injections with needles that have been used with infected people. Current interventions for malaria are not producing the necessary reductions in incidence and mortality (Editorial "Developing" 2007). There are plans for an effective malaria vaccine by the year 2025.

The reappearance of human tuberculosis has resulted from ineffective *suppression of the bacteria* within the human body. One-third of the world's population is infected, and there are 2 million deaths a year from tuberculosis (Floto et al. 2006). The current causative agent for human tuberculosis is a clinically important mutant strain of *Mycobacterium tuberculosis* that has become multidrug resistant and globally epidemic prone. This strain is resistant to rifampin, isoniazid, and the combination of isoniazid and streptomycin.

A T-cell response to the bacteria results in hallmark granulomas that limit mycobacterial replication and control immunopatholological consequences (Floto et al. 2006). There is balance between appropriate response to microbes and the inducement of immunopathology. The extensively drug-resistant tuberculosis (XDR TB), a multidrug resistance possibly acquired through inappropriate therapy, caused an outbreak in South Africa. It was stopped by rapid intervention, accelerated diagnosis, and a full spectrum of tests for all antibiotics at the outset (van Helden et al. 2006). The public health lesson is to use rapid intervention to prevent transmission.

Drug-resistant bacterial strains, having little growth fitness costs and no compensatory genetic amelioration, are more likely to spread in the human population until, at least temporarily, another treatment drug proves effective, low cost, and without serious side effects. In general, multidrug-resistant strains of bacteria do threaten global disease control efforts and can create a significant public health burden.

There have been many repeated *E. coli* outbreaks traced to contaminated fruits and vegetables. Proposals to irradiate, with gamma rays, all produce, poultry, and meat have been resisted. Irradiation kills most insects and bacteria, so foodborne disease could be substantially reduced. The FDA permits irradiation of meats but requires warning labels that suggest a risk. Proponents of irradiation suggest that there can never be enough inspectors to track all the sources of fresh produce ("*E. coli*" 2006). It is a political question since opponents of irradiation advocate better source and distribution inspections, quality assurance, and testing programs.

Similar opposition has occurred for HIV/AIDS (human immunodeficiency virus/acquired immunodeficiency syndrome) testing. A Chinese program includes mandatory testing among prisoners and government workers under a general health examination worker consent procedure (Mills and Rennie 2006). Proponents of testing state that community protection and public health outweigh individual rights, particularly in high-risk areas or for drug users. Opponents stress individual responsibility, individual human rights, and the need for informed consent and "opt-out" provisions in either routine or mandatory test programs. About 25% of the 850,000 infected persons in the United States are unaware that they are HIV positive. They can innocently infect others, particularly careless health care workers.

Just before the Christmas holiday in 2006, a hospital in Los Angeles, California, closed its neonatal and pediatrics intensive care units (ICUs) to new admissions. A potentially fatal bacterium infected seven children, including two who died. The *Pseudomonas aeruginosa* bacteria had been detected on November 30 in the neonatal unit, and on December 4 it was shut down.

Later, it was detected in the pediatrics ICU, which was shut down December 15. The source of the outbreak was a contaminated laryngoscope, the medical instrument used to examine an infant's larynx. This information is from a news release that is typical of those published in local cities throughout the country. The *Pseudomonas* organisms are frequently found in hospital sinks, urine receptacles, and sometimes antiseptic solutions.

An investigation of the death of the two premature infants was conducted by the County Department of Public Health. Previously, the laryngoscope blades had been sent to a central in-house sterilization unit at the hospital. The purpose of a *central cleaning unit* is to ensure that all soiled or contaminated instruments are correctly sterilized for use according to the instructions prepared by the device manufacturer. Earlier in the year, prior to the deaths, the hospital policy of sending the blades to the central cleaning unit was changed (Chang 2006). The cleaning was then done by the respiratory staff. Although the procedure actually used was unknown, it is a common practice for nurses just to wipe down such instruments after use since they are not instrument cleaning specialists. About 10% of the 2 million hospital-acquired infections each year are caused by *P. aeruginosa*.

The actual spread of infectious diseases among the general population can be quite high. The percentage of women in the United States who have the genital herpes virus was 25.2% in 1988–94 ("Baby Talk" 2006). The virus may be inactive for long periods of time, producing no symptoms, and the majority of women who are infected may be unaware of the fact. Only 1% to 2% of pregnant women who are infected take drugs for herpes. Antibodies accumulate and are passed on to the baby, thus protecting the infant. But, some women pass on the virus to their babies if the mother was newly infected during the last trimester of the pregnancy. The overall potential of harm is such that screening of asymptomatic women may not be warranted on a drug side effect and cost-benefit basis. Medication and cesarean section delivery to prevent neonatal herpes should occur if there is a genital herpes outbreak during pregnancy. In essence, many people may have an infectious disease and be asymptomatic, and health care workers could become exposed from an ostensibly clean patient.

Herpes simplex virus-1 (HSV-1) infects about 80% of young adults worldwide (Casrouge et al. 2006). Encephalitis is a complication of HSV-1, with a mortality of about 70% unless treated with the antiviral drugs vidarabine or acyclovir. Despite treatment with the antiviral drugs, there can be neurological sequelae. The pathogenesis of the HSV-1 encephalitis may be from a monogenetic trait that results in immunity impairment that is pathogen specific. That is, in a small minority of infected HSV-1 individuals, apparently healthy, there is an immunodeficiency that results in a specific susceptibility to encephalitis. This contrasts with gene lesions that confer vulnerability to multiple common infections, that is, a generally impaired antiviral response from a polygenic predisposition. There may be other infectious diseases that reflect a monogenetic immunity disorder, unlike most primary immune deficiencies, that may benefit from more specific treatment options.

There is a higher risk for getting infected and spreading disease by medical personnel in close contact with patients, particularly the elderly, frail, and vulnerable. Health care workers have a special duty to protect themselves and others against the spread of infectious disease.

Special attention should be given to illegal immigrants, tourists, and visitors from foreign countries that have a high incidence of bloodborne diseases such as the hepatitis A, B, C, D, E, and G viruses, which cause liver damage and occasional epidemics. Also on the watch list should be HIV, which destroys lymphocytes and causes AIDS, which affected 20 million people worldwide in 1996, and the epidemic has been increasing. Transmission is from body fluids, including blood, semen, vaginal secretions, cerebrospinal fluid, breast milk, and possibly tears, urine, and saliva. Bacterial infections are common, such as by the tuberculosis bacilli, which affects the pulmonary system, the brain, joints, intestines, and bones. Also on the watch list should be fairly common transmissible herpes virus infections such as herpes simplex (HSV-1 and HSV-2), herpes zoster, and Epstein-Barr; these are well known by the general public.

About 28,000 patients die each year from blood infections caused by central-line catheter procedures (Seward 2006). These catheters deliver drugs and nutrition through a vein in the neck, chest, or groin. In a study of catheter-related infections in ICUs, where the hospitals agreed to use basic patient safety procedures, the infections were practically eliminated. The safety precautions were that doctors and nurses wash their hands, wear protective clothing, disinfect the catheter site, remove unnecessary catheters more quickly, avoid placement of catheters in the femoral vein, and call attention to the infection problem. Patients in ICUs spend about half their time with central-line catheters in place. The treatment costs for blood infections average $45,000. The lead investigator for the study indicated that such infections could be "practically eliminated."

There are universal precautions to control the spread of infectious disease, such as hand washing between procedures, cleaning instruments between patient use, using personal protective equipment, and using working techniques such as avoidance of health care worker cuts from sharp edges, bone spicules, and needle punctures. Some hospitals utilize *universal surveillance*, which consists of testing every patient, at admission, for possible infection. A positive test results in patient isolation and administration of appropriate antibiotics; *all* people entering the patient's room must wear gowns and gloves. Some hospitals request that prospective patients be tested at the referring doctor's offices, perhaps a week before admission, so that the hospital staff can be alerted. About two-thirds of all health care infections involve antibiotic-resistant strains, and even vancomycin may not be effective. There are countries, such as The Netherlands, where hospital-acquired infections are rare. However, we should expect and be prepared for new infectious disease challenges, resultant precautions, and public fears.

Emergency Services

There have been well-known difficulties in providing emergency care because of costs and limited facilities. Some hospitals have two emergency entrances, one for ambulances and another for walk-ins who are less likely to be able to fund their treatments. It is not unusual to see a person in severe pain, perhaps from a broken arm, waiting for several hours in the walk-in reception room after checking in. Emergency rooms are often overcrowded. This discourages prompt treatment and engenders bad attitudes that may result in an indifference to the pain of others. Recently, a coroner's jury ruled that there was a criminal homicide in the death of a woman who had a heart attack after waiting for 2 hours in an ER. These errors of omission stand in stark contrast to the broad skills, high talent, and personal dedication of most overtaxed ER specialists.

The California Hospital Association in 2006 stated that hospitals provide emergency service to all patients irrespective of their ability to pay. Also that one of every five Californians are uninsured, that $7 billion in care was not paid for in 2005, and that 46% of all California hospitals operated in the red. (www.calhospital.org, accessed July 2007).

For incoming ambulance patients, the ER staff may have difficulty deciphering messages because of radio chatter and other communication problems. An incorrectly heard voice message may lead to an incorrect diagnosis. To avoid such medical errors, clear communications equipment may be needed, such as throat microphones, ear speakers, push-to-talk radio buttons, and belt-mounted voice amplifiers. This may ensure hands-free operations, less voice strain, open radio frequencies for backup, and a reduction in radio traffic. There are new requirements for disasters, such as portable decontamination tents, air-purifying respirator suits, and better means of communication between staff and between staff and patients. Disasters include hurricanes, chemical spills, airborne infections, and release of nuclear or biological agents. There should be practiced designated emergency response teams. Translators should be available for foreign language patients. The more confusion and chaos there is, the more errors are expected.

Automatic alarms or functions may be triggered by key word phrases in a voice-activated portable computer worn by emergency personnel or hospital staff. This is another hands-free instant message service that saves critical time in rendering medical services, particularly in confusing or chaotic situations.

Pediatric emergency care often suffers from a lack of equipment designed for infants, children, and adolescents. This includes needles, catheters, breathing tubes, and oxygen masks in different sizes. Similar variables have resulted in a 10% medication prescribing error. The pediatric ER should have pediatric crash carts with special kits coded for patient weight, size, and age.

Reduction of medical errors requires trained staff and proper equipment. There should be separate waiting rooms to avoid contact with sick adults, a capability for isolating examination and treatment rooms, and comfort items such as animated DVDs, video games, and wireless Internet access. Pediatric emergencies include sports injuries, asthma attacks, bone fractures, head trauma, seizures, dehydration from diarrhea, and preparations for mass casualties.

Because a child is receiving emergency medical care, parents and relatives may display excessive emotional and illogical attitudes. Patients may have special childhood problems such as foreign objects stuck in the throat, respiratory tract, or other body openings. These objects may also include food, bones, nuts, or vegetables not properly chewed. Convulsions may be caused by a variety of unrelated conditions (Henderson 1973). There may be mild cerebral anoxia from breath holding. To avoid error, specialists in pediatrics and pediatric emergency medicine should be available, and ER staff should attend seminars and follow the recommendations of the American Academy of Pediatrics, American College of Emergency Physicians, and the Emergency Nurses Association.

Handovers and Interactions

A frequent source of medical error has been called *handover* or *handoff error*. For example, a tired nurse completing a work shift attempts to communicate to the incoming, new, or next-shift nurse to convey or transfer what seems to be important information about the patients. The incoming nurse listens and attempts to understand and act on that information. There may be errors in what is communicated, how it is communicated, how it is interpreted by the recipient, and what to do about it. In a handover to the next shift, both nurses process the information by their brain's verbal comprehension, past and immediate memory, and executive function domains in conjunction with other associative areas. In any human social interaction, there are mental evaluations of social, occupational, and cost consequences. Standardized methods of team communication during nurse handoffs reduce communication errors (Hohenhaus et al. 2006).

Patient handoffs include those during nursing shift changes, transfer of complete patient responsibility from one physician to another, assignment of temporary responsibility when staff leaves a unit for a short time, transfer of on-call responsibilities, transfer from emergency to inpatient units, and transfer between hospitals, nursing homes, and home care.

To reduce medical error, the information should be communicated accurately, directly, and clearly and with the degree of importance evident. By using written charts, logs, computers, laptops, personal assistant devices, or even coded Internet messages, such communications may be easier to formulate and deliver, capable of revisitation, and may constitute a truthful historical record. Currently, handoff miscues remain a major source of medical error (Landro 2006c).

In manufacturing industries, including medical device companies, shift handover incidents and accidents are commonplace. For example, in the 2005 Texas City fire that killed 15 people, the night shift workers filled a petroleum isomerization column, and the next day shift continued to fill it until it overflowed (Nimmo 2006). The fluid quickly vaporized, a cloud formed, and then it was explosively ignited. The error was in the communications between the night and the day shifts, a handover error. The error countermeasures included a shift log book with defined and scheduled data collection entries and the indication of the status of all operations. Electronic log books open the information to all, including management. Remedial training may be necessary to effectively document events, conditions, and monitoring needs to avoid unexpected problems. A second tool is formal shift handover meetings to identify the status of problems, potential threats, and plans. The meeting should be short and have a focus on what is important during the transfer to a new shift. The third tool is a shift team meeting early in the shift to share information, review the progress of the earlier shift, and deal with broader issues. The fourth tool is a shift-monitoring plan to identify unique work routines or locally devised procedures. The purpose is to provide guidance on which procedures are desirable from a management perspective. This provides an opportunity for implementing a continual improvement objective. Such tools constitute a proactive approach to reducing unwanted error.

Whenever there is an interaction between humans and machines, various equipment items, professional disciplines, or hospital departments, there can be communication errors. For example, with the advent of conscious sedation using drugs such as fentanyl and Versed, most surgical procedures are performed in hospital outpatient departments, surgical centers, and physician offices. There are not enough anesthesiologists and nurse anesthetists, so health care providers seek the right to administer their own sedation drugs. They may not be adequately trained in another discipline, so the risk of errors and complications increases. These problems include cardiac arrest, respiratory complications, inadequate sedation, starting surgery before the drugs take full effect, overdoses, and death.

Preventing injuries from bioterrorism involves performance in a different context and a resulting higher risk of errors from the unfamiliar, new, and unexpected. The high-priority agents include anthrax, botulism, smallpox, and pandemic influenza; these spread quickly and have high mortality rates. There are also food safety threats such as *E. coli* and salmonella, and potential diseases, such as from the hantavirus. There are special treatments and vaccines that are unfamiliar and carry unique risks and error potentials.

If there is a likelihood of handover and interaction errors, then an error management program should have a hotline for reporting errors, near-misses, and conditions that foster errors. Recommendation for countermeasures should also be encouraged. The hotline reports of errors should remain voluntary and confidential.

Health Care Facilities

Cost Benefits

There have been many discussions on the disparity between health care performance outcomes and the investment now required to operate the system (approximately 16% of the gross domestic product of the United States). There is an expressed need for more and better preventive medicine and direct public health actions. Health care insurers cite the high costs of preventable medical error. The number of uninsured or partially insured patients is intolerably high. Comparisons are made with health care operations in other countries. These discussions seem to infer or conclude that fundamental changes are needed in an activity already beset by ongoing major transitions. All of this has acted to focus attention on hospital design, layout, functionality, and cost-effectiveness. Could better architectural design reduce medical mishaps and health care costs?

Clean Air

Hospitals should have clean indoor air, but most air conditioning systems recirculate or recycle a significant proportion of the air to reduce operating costs. The recycled air transports contaminants (such as airborne microbes) from one area to all of the areas serviced by the ventilation or air conditioning unit. It may introduce outside air contaminated with particulate matter, toxic substances, combustion products, chemical emissions, and other sick building pollutants. The design of the HVAC (heating, ventilation, and air conditioning) should control moisture levels and clean the incoming outdoor air. There should be no recycling of air that could spread infection.

There may be a misunderstanding of what is required for clean air. Air filtering is accomplished by methods of interception, impaction, and electrostatic deposition. Air cleaning removes particulate matter and gaseous contaminants with substances such as charcoal and potassium permanganate. Volatile organic compound (VOC) removal is accomplished by air ionization, ultraviolet photocatalytic oxidation, and gas/vapor sorption. There are high-efficiency particulate arrestance (HEPA) filters that are 99.97% efficient in the size (about 0.3 micrometers in diameter) of particles that penetrate deep into the respiratory tract when inhaled. The actual HEPA measured efficiency may be 50% to 80% (Chen et al. 2006). The outdoor air intakes should be at a high level of the building, away from vehicle emissions and dust swirls. For these often-complex situations, there are often errors by those who adjust, maintain, service, and repair such HVAC units.

One part of the HVAC system is a smoke control system to prevent smoke migration from a fire origin through the ventilation duct system to other building sites. The air supply and exhaust system should automatically vent smoke and combustion products and prevent smoke from a surgical suite from entering the return intake ducts. Nonflammable inhalation of anesthetic agents, such as nitrous oxide, should be used for relative analgesia or conscious sedation. There should be positive-pressure stair towers to exclude smoke. Smoke consists of airborne solids, liquid particulates, and combustion gases. There may be problems in locating the mechanical equipment rooms, including HVAC units, in high-rise or multilevel hospitals with different patient care functions on each floor. This short summary of fire and smoke control requirements should suggest that there are error-prone problems if such systems are not carefully designed for regulatory compliance, operation, maintenance, and repair.

Building-related diseases include legionnaire's disease, Pontiac fever, hypersensitivity pneumonitis, humidity fever, and lung cancer from radon or tobacco smoke. Probable diseases include tuberculosis, influenza, and common colds (Watson et al. 1997, p. 93). Architects have design countermeasures for each, but much relies on human maintenance and error avoidance.

Clustering and Flexibility

Each hospital function and department should be located and arranged for maximum efficiency, timing, quick response, and health care delivery. This clustering or subgrouping will vary by hospital and will change over time. The design of the hospital should accommodate staff function clustering and the changes that will occur in the future. An example of the needed architectural flexibility is the change from a single portal for emergency services to two portals (ambulance and walk-in), then the addition of "urgent care" facilities as the old emergency services were overcome by people using the ER for general services.

As new imaging equipment, computer electronics, and other advances occur in the hospital, there will be building changes. Smart equipment should fit into a smart hospital without constant redesign.

Patient Variability

There is great variation in the patient population, including the size of persons. Size matters for the overweight or obese, some 60 million people or 30% of the general population. Appropriate patient lift systems, whether mounted in the patient's room or portable, are necessary to prevent the all-too-common overexertion and back injuries of health care workers. The lifts should be capable of lifting 600 to 1,000 pounds.

Doorway widths of up to 48 inches and bathroom doors with a 60-inch width can help two nurses to assist in patient access. Wide chairs, up to 40 inches, should be able to support up to 750 pounds of static load. Heavy-duty shower grab rails should be able to handle a 500-pound load and be placed all around the periphery when the patient is standing and at a lower level when the patient is sitting and starting to stand. Beds can be equipped with weight scales with readings up to 1,000 pounds. There are beds that can expand in width from 40 to 50 inches or 37 to 54 inches. There are motor-driven beds for patient handling. The placement of toilets should provide for access by grab rails and sideways movement rather than twisting movements for postsurgical patients. Storage areas should be able to accommodate oversize gurneys, beds, and wheelchairs. The basic rules are that larger dimensions will generally accommodate patients of smaller size and that improper size facilities invite human error.

General Criteria

The hospital should have soundproof walls for noise reduction and privacy. The walls should not be covered with substances that attract mold (such as vinyl wall coverings). Walls of different color or decoration would help improve patient orientation and reduce the possibility of patients getting lost. The floors should be slipproof and easily cleaned, and all corners of the wall should be rounded. Glass should be used so the patients are always in view of the nurse, with glass alcoves for computer entry and readout and sliding drawers for medication delivery to reduce the infection risks from repeated entry of the patient's room by nurses (who wear the same clothing as they go from one room to another).

If the line of sight from the nurses' station into the patient's room is some-how incomplete, partially blocked, or obstructed completely, then a sup-plementary vision system should be considered. In lieu of direct vision, a closed-circuit television system could be used, with a low-cost unobtrusive camera located in or just outside the patient's room and a visual display at the nurses' station. The camera could have low light and zoom capability. Two-way speakers could be utilized. Motion sensors could be used to detect patient movement. The sound of medical equipment alarms could be relayed by the speaker system to the nurses' station as desired. There are systems for the display of images from multiple cameras that have been used for build-ing security for several decades, have been greatly improved over the years, and can be adapted to local preferences.

A digital video system may permit the detection of smoke long before a smoke detector signals a potential fire. It could provide a corridor security function and be integrated into overall building intruder surveillance, coor-dinated emergency evacuation plans, electrical standby systems, and fire sprinkler systems. However, with new integrated complex systems, there is an increased likelihood of error during the initial installation, maintenance, and operation.

There should be uniform natural-like or daylight lighting for offices, corridors, and nurses' stations to avoid errors in reading, squinting, computation, or visual sighting tasks. Subdued or adjustable lighting may be more appropriate for the patients' rooms. The size and shape of each patient room should be identical so that physicians and nurses can quickly find emergency oxygen lines, syringes, and other items in a standardized format and location (Naik 2006).

Hospital beds should be designed and located to reduce the risk of entrapment (when a part of the patient's body becomes caught between parts of the bed). The FDA received 700 reports of hospital bed entrapment during the past decade. Patients who are elderly, frail, confused, or have uncontrolled body movements are at a higher risk for entrapment.

Heater vents should be located above the windows to reduce the condensation that attracts fungi. Showers should have strong grab bars to help stabilize entry, standing, sitting, and exit. Doors should have hospital pulls to permit entry using wrist, arm, or forearm movement when a caregiver's hands are occupied. Doors should be capable of resisting abuse from wheelchairs, rolling beds, stretchers, and medication carts. Hospital-type elevators should be of adequate capability and size without the appearance and function of freight elevators. All hardware should be compatible with the limited capabilities of the aged, infirm, sick, and disabled.

A hospital building failure (diminished functionality) may occur if there is an error in the emergency power system. This is the system used during power outages for emergency communication, egress, patient food service, patient heating and cooling, life support systems, and some central sterile equipment. For life safety purposes, the Joint Commission on Accreditation of Healthcare Organizations mandates the use of National Fire Protection Association standards 70, 99, and 110. An inquiry should be made regarding the electrical power service in emergencies that relates to patient safety. Each facility initially determines its individual needs. What happens when the hospital is cut off from utility power during a catastrophic event? Are there new needs for the electrical distribution system, its maintenance and operation plan, and the servicing of such backup power systems?

Limitations of Service

Despite the excellent training and professional demeanor found in the medical specialties, there is a gap between the ideal and the constraints of reality. Some hospitals experience their own quickly spreading epidemics of infectious diseases despite good intentions and due care. Despite an increasing population, there has been a reduction of ERs. Of a general population of 300 million people in the United States, some 46 million (about 15%) have no health insurance. No insurance may result in the denial of treatment or a lesser form of treatment. Health insurance premiums may consume most of the wages of unskilled labor. A lack of early public health services may increase later demands for treatment. Comparisons have been made with other countries, and medical services in this country may cost more, sometimes twice as much per capita. There are problems in dealing with the HMO bureaucracy, insurance payment limitations, and rapid advances in methods of treatment. Despite these limitations, there has been steady progress and improved medical services. Criticism is to be expected during periods of rapid transition. The ideal will become the reality.

Caveats

Proof

Much of medical error prevention utilized in medical service operations is based on simplistic theoretical assumptions that have not been proven by controlled tests and acceptable experimental methodology. Unproven and indirect generalizations are used in the hope that they might be effective to some degree.

Specificity

The focus for error correction and prevention should be on specifics, evidence-based facts, proportional causation, and tailored countermeasures. One objective of specificity is the identification and development of detailed concepts and remedies capable of universal understanding and implementation.

Culture

Reduction of the spread of nosocomial (hospital-acquired) infections to patients and staff requires a strong commitment in the form of developing a culture of infection protection. Mandatory immunization of medical workers is desirable.

Architecture

There are many modifications that could be made in building design that would reduce medical errors relating to infection control, computer services, call signal responses, patient movement, privacy, patient visibility to nurses, fire and smoke control, air filtering, and other current problems prevalent in outdated facilities. Touch surfaces should be antimicrobial.

Handoffs

There are standardized methods of communication that can reduce shift handover errors, miscues, patient safety problems, and medical errors. Additional human factors research is desirable.

Emergency Services

Provision for emergency services is costly but necessary. Special attention should be given to advances in technology that might be helpful.

Call Lights

Lifeline calls should be distinguished from housekeeping and personal care communications. Protocols should be developed for rounds in accordance with need, staff availability, and error reduction efforts.

Basic Hygiene

The extension to more routine activities of basic safety and hygiene practices could dramatically reduce life-threatening infections. Simple infection control measures are far more important than many health care workers care to believe as they continue customary practices.

Mutations

In 1999, one of five hospital patients developed a bacterial or viral infection, and 90,000 died (Centers for Disease Control and Prevention [CDC] statistics). The world's first antibiotic was penicillin, introduced in the early 1940s; it blocked replication of the cell-wall-building process of *S. aureus*. Within 2 years, the bacterium developed resistance to penicillin by changing one gene so that the antibiotic could no longer attach to the cell wall. Vancomycin works in a similar manner, but *S. aureus* had to mutate all five genes that the drug targets. This took longer, but vancomycin-resistant *S. aureus* is now commonplace. *Staphylococcus aureus* and other infectious agents have proven resistant to the penicillins, then methicillin, then vancomycin, and other anti-infective drugs. There are now varying degrees of resistance to many antimicrobial drugs that are used for treatment or prevention in humans and animals. Drug resistance may occur from mutations that can be passed from strain to strain (gene swapping). The greater the use of an antibiotic (amount and pattern), the greater the opportunity for mutations will be, so a lack of prudent drug use may quickly reduce its effectiveness. Medical specialists have been repeatedly warned about the adverse effects of overuse (nonprudent use), which could be considered a medical error, but overuse may continue because of a proven therapeutic effectiveness or the patient's desire for such drugs and their beneficial effects. The disregard (error) of bacterial mutations (evolution) may rest on beliefs or hopes that the glycosyl-transferase enzymes that are involved in bacterium cell wall biosynthesis provide accessible target sites for new drugs. The bacterial membrane protein has no mammalian equivalents, so new antimicrobial agents should have few side effects. If significant overdosing of a drug continues, then there may be questions why, and the answers may depend on who is asked. The question is of interest to human factors specialists because of their research on human error, its causes, and its prevention.

Decision Making

Physicians have been taught to match a patient's symptoms to their mental model or learned encyclopedia of diagnosis and treatment. The tendency to cherry-pick symptoms and use rules of thumb derived from prior cases has resulted in misdiagnoses (medical errors) about 10% to 15% of the time. There should be an awareness of this form of bias, uncertainty, and problems in routine decision making. Any patient can be atypical, and the examination should probe accordingly.

Surgical Infection Control

During surgical procedures, there is a significant patient risk from airborne and contact contamination. Bacterial infection may require prolonged use of antibiotics, require more surgical procedures, and escalate costs. In addition to asepsis procedures, there is usually prophylactic administration of antibiotics before and after surgery; the use of filtered laminar flow ventilation, body exhaust suits, water-repellent drapes; special attention to comorbidities and immune suppression; and reduced-time minimally invasive procedures. Unfortunately, where infection remains a serious problem, the surgical infection control program is often incomplete or not enforced.

Choice Behavior

The importance and complexity of human behavior in avoidable disease is suggested by the extent of cigarette smoking in the United States. Smoking is a voluntarily initiated or choice behavior by an individual. In 2005, there were 45.1 million smokers in the United States (CDC statistics). Smoking was a causal risk factor in 90% of all male cancer deaths and in 80% of female cancer deaths and posed an increased risk of 1.3 for those exposed to second-hand smoke. The cancers are small-cell carcinoma, squamous cell carcinoma, adenocarcinoma, large-cell carcinoma, and eight other types of tumors. The self-destructive nature of the nicotine addiction and the difficulty of stopping the behavior is well known and highly publicized, but it continues. Modifying certain types of human behavior could be very difficult.

Rules of the Road

There are some human behaviors that could be easily modified. An analogy is the automobile driver who drives within marked traffic lanes, below posted speed limits, to defined parking spots.

The driver obeys traffic signals, signs, and other indicators of the rules of the road. This is a socially required set of behaviors, with enforcement of the traffic rules, and the rules have a high compliance rate. The vehicle traffic control system evolved over a long period of time with gradual changes in the traffic mix and environment. In this book, there are many medical errors identified and many remedies suggested . Some may have immediate high compliance, some may have evolving compliance, and others may be difficult to implement. This is one reason why the entire contents of this book are important and should be considered.

The User Process

The hospital architecture should be designed to accommodate the caregiving processes that could reduce medical error. Nurses and physicians should not be forced to "work around" the built environment or "deal with" many variations in unit layout. The hospital should be designed for them to achieve uniform practices and high levels of performance. There should be standardization of room size, type, and layout, with similar medical gas drops, oxygen outlets, medical equipment, drawers for bandages and swabs, headwalls, and bed locations against the same wall. Bathrooms could be located on an outside wall to improve nurse's sight lines into the bathroom for monitoring. There should be handrails running from the bathroom to the bed to prevent falls. Decentralizing nurses' stations, for example, may involve having 12 beds in a circular orientation around a single nursing station so that nurses are always within 20 feet of all patient rooms with clear sight lines to each. A central nursing station and workroom preserves the camaraderie and social collaboration of the caregivers' team effort. If treatment costs are high, such as in proton therapy for cancer, there must be a focus on equipment utilization, neutron shielding, patient setup, and dealing with support personnel. The objective is to achieve the best user process for the facility.

4

Medical Devices

The medical device industry has been undergoing rapid growth, diversification, and increasing technical sophistication. Its unique products and services already have gained a fairly good reputation and have demonstrated substantial proven value to society as a whole. The future certainly promises even more creative diagnostic and treatment systems that should result in significant public health benefits. However, the industry also has unique risks and vulnerabilities, as described in this chapter, that can and should be appropriately managed. It is not a fully mature industry, and special precautions should be taken in the selection of such devices. This chapter emphasizes the role of the medical device manufacturer and how that impacts other medical services.

Risks

Inherent

Medical devices may be located on the surface of or inside the human body. In that direct contact position, even a minor malfunction or failure might cause substantial injury to a body structure, organ function, or fluid component. A device's failure to perform as expected might result in immediate irreparable damage to the device and to the patient. Even a low-frequency adverse event or malfunction may constitute an unacceptable risk because of the severity of the possible consequences. For medical devices, a very high standard of care in design, manufacture, implantation, robustness, and design safety is essential given such risks.

Illustrative Problems

After surgery, to close a wound for healing, a variety of sutures have been used. Nonabsorbable sutures composed of cotton, black silk, or nylon were used to sew up the wound, although they could harbor infectious organisms. Absorbable plain or chromic gut sutures, made from sheep intestines, might create a foreign body reaction. Mechanical staples, made of a steel alloy, were fast but required stitch removal and left scarring. The latest medical device is a mechanical skin stapler that uses absorbable skin staples to close wounds in a process that can save up to 2 hours. In other words, there was a gradual improvement in suture materials and stapling devices. The problems were considered minor compared with the beneficial results, assuming good professional discretion regarding the most appropriate wound-healing method in a given situation.

There were problems with some implanted cardiac defibrillators, which were predisposed to short circuits (arcing). The prime risk was failure to deliver the intended therapy when needed because of a defect that could result in death (Myerburg et al. 2006). When this occurred in 2002 to mid-2005, there was criticism that there was a failure to effectively communicate the life-threatening nature of the malfunctions to physicians and patients. It was assumed that corporate cultures create employee loyalties to corporate goals, resulting in unintended bias and distorted perceptions about product performance. It was stated that medical devices cannot be entirely free of design and manufacturing defects, that knowledge of product design features constituted informed consent, and that the patient should obtain and understand the necessary product information. This was considered an engineering problem, not a medical problem. In essence, given a defect, there are advocates for all conceivable sides of the issue. The prime objective should be to minimize, to near-zero levels, the possibility of high-risk problems by instituting appropriate product development controls as described in this book.

Defibrillators shock an abnormal heartbeat into a normal pattern or rhythm, whereas pacemakers correct a slow heartbeat. In 2006, one manufacturer recalled nearly 50,000 pacemakers and defibrillators, about 27,200 of which were implanted at time of recall. About 200,000 physician warnings and recalls had been previously issued (Westphal 2006a). The pacemaker malfunctions could result in syncope, a temporary loss of consciousness. The malfunction was caused by a low-voltage capacitor supplied by a vendor. This emphasizes the importance of monitoring the incoming quality of vendor-supplied components, including final testing of the assembled device.

In December 2005, the Food and Drug Administration (FDA) announced that there were 67 clinical failures, including 7 deaths, involving three models of the defibrillators (Burtka 2006). Subsequently, there was unfavorable publicity in newspapers and other media.

There are sales in excess of $5 billion for wire mesh tubes (stents) used to keep open human arteries that were previously clogged. Cardiac stents may be drug coated to prevent tissue growth or reclogging while inserted in the artery. News of a slightly elevated risk of late stent thrombosis (blood clots) in drug-coated stents as contrasted with bare metal stents resulted in cardiologists extending the use of anticlotting medications (Westphal 2006a). This should involve a stent label change for the use of anticlotting medications (such as Plavix) beyond the normal 3- to 6-month time period, although the early FDA decision was that no label change was required. Stents are also in use for purposes beyond their approved use. This emphasizes the need for postmarketing surveillance of medical devices and taking actions beyond those required by U.S. and European regulators.

It is unfortunate that when problems arise with a medical device, the FDA and the manufacturers quickly say "these devices are safe and effective when used as indicated" or when used "for the treatment of the type of cases for which they were approved." This may serve to disarm those who should take added precautions in the selection, application, and use of the devices. It is also important because such devices are often used for purposes unapproved or exempted by the FDA and that already have an uncertain risk.

The world's first self-contained artificial heart received limited FDA approval, in 2006, in the form of a "humanitarian device exemption" (Westphal 2006b). The limitation was that it be used only for patients with advanced heart failure who are not eligible for heart transplant or other alternative treatment. The artificial heart replaces the diseased heart's pumping action. It has a rechargeable internal battery and a microprocessor implanted in the patient's abdomen. Patient trials and product improvement began in 2001. The life extension of the patient is above 5 months, and the service life of the device is 18 months. The cost is about $250,000. What is instructive is the risk-benefit analysis, in which the patient's 5-month life extension (the benefit) was balanced against the risks (stroke, bleeding, and the adverse effects of anticlotting medications).

Human Error

Human errors are generally unwanted events, even if detectable and correctable, considered to be of only nuisance value, or simply a manifestation of a condition that cannot be made more user friendly. There are several general independent sources of harmful error. It may be *medical error,* such as if there is an implantable device, because surgical intervention always poses a unique class of hazards and potential risks. It may be *user error,* such as if there is a handheld wireless remote control or programmer that is abused or incorrectly used by the patient. It may be an *experimental error* since product development almost always involves unknowns and uncertainties that may not be appropriately controlled. It may be an *environmentally caused error* associated with devices immersed in hostile surroundings of corrosive bodily fluids, an immune process that recognizes the device as a foreign object, external electromagnetic fields that induce unexpected signals in electronic circuits, and anatomical (body) movements that stress or displace sensors, leads, or control components. Each of these errors may have a trigger or contributing cause, such as abusive care by the user. In fact, there are many sources of error, with associated risk, that should be controlled by appropriate error-reduction techniques and methodology. The magnitude and diversity of the risks of injury are unique to the medical device industry and are enhanced by vigorous exploration and experimentation at the expanding boundaries of microelectronics, micromechanics, microfluidics, microfabrication, and creative design in areas of emerging human need.

Neurostimulation

An example of the kind of risks that may be undertaken with medical devices is the use of electrical neurostimulators or leads located adjacent to the spinal cord for pain management and relief, lead contacts at the vagus nerve in attempts to control depression, or leads on or in the brain for diagnostic or treatment purposes. During nearly 40 years of neurostimulator experience, thousands of such devices have been successfully used. There are a wide variety of implants in use, ranging from tens of thousands of cochlear implants to improve hearing to glucose monitoring and dispensing systems. The demand for various treatment options encourages developmental risk taking, but the risks should be evaluated, assessed, and controlled whether the risk is from the design, fabrication, or errors by others. This requires considerable care and oversight.

The market potential for neurostimulators is large. At least 80% of the general population will experience some form of low back pain during their lifetime (Dutton 2004). A small percentage progress to chronic disabling pain, and this translates to millions of people who suffer chronic functional benign impairment with episodes of severe pain. The pain may be from shingles (herpes zoster) caused by a varicella-zoster virus infection that first results in chicken pox and then enters the ganglia of spinal or cranial nerves. A so-called slipped disc or bulge (herniated intervertebral disc) may press against a nerve root (root pathology), and a radiating sharp pain down the leg or sciatica may result. There are many causes of pain, and pain relief may be actively sought by many patients. A neurostimulator and the associated surgery, in 2006, cost between $20,000 and $30,000, which is generally covered by insurance. There may be other surgical costs to relocate leads (wires) that move, to replace implanted batteries (after 5 to 10 years), and to explant and replace the device after expiration of its service life.

There are pain management centers that provide a variety of services for patients suffering pain. They may first elect noninvasive treatments such as pain-relieving medications in combination with physical therapy and psychological support. More aggressive therapies include cryoanalgesia for nerves near the surface of the skin, implantable analgesic drug delivery into the fluid surrounding the spinal cord, nerve blocks injected near specific spinal nerves, and radio-frequency (heat) ablation of a nerve. As a last resort, spinal cord stimulators (neurostimulators) are used to apply a small electrical current into the epidural space of the spinal cord to interrupt the transmission of pain signals. There are many options, a trial-and-error process often results, and therapeutic failures are common due to errors on the part of the therapist or patient.

Corrective Remedies

Often, the surgeon may be blamed for committing a medical error when an implantable medical device is positioned at the wrong site, using the wrong procedure, and on the wrong patient. Such stories abound. Less may be said about effective corrective remedies, including procedurally correct instructive videos, marking the patient by one person, checking the patient's identification wrist bracelet by another person, reviewing medical records by still another person, and perhaps employing radio-frequency identification and location devices. The real objective should be preventive action rather than after-the-fact corrective responses.

About 20% of people have a hole in their heart, and some need an umbrella-shaped implant to seal the hole (Neergaard 2006). The flaplike opening between the two upper chambers of the heart does not grow shut during infancy, causing a defect known as patent foramen ovale. It may cause a severe migraine with an aura and an increased probability of a stroke. The implant procedure involves inserting a balloon catheter that is threaded from the groin to the heart opening; the balloon is inflated and leaves behind an implant to seal the hole, and then heart tissue forms on the mesh, creating new tissue. The procedure costs, in 2006, were about $23,000. Thousands of the devices had been implanted by 2006 when the FDA stopped promotion of the devices pending final proof of their clinical effectiveness. The devices had been implanted under a special FDA rule for promising devices for rare medical conditions. The withdrawal of FDA special approval was on the basis that the conditions are no longer rare. There have been some human safety tests on dissolvable implants and tests using radio-frequency energy to weld the holes shut. The corrective remedy, stopping promotion of the devices, was tardy, and their implantation continues unabated.

Product Life

Medical devices have special problems related to the risks of long-term use in the hostile environment of bodily fluids. The service or product life of a medical device varies regarding its material composition, design, and function in the human body. The problems relate to the limited ability to test a product in humans during its design phase and the fact that postimplantation monitoring is almost akin to experimenting on humans. A recent vertebral artificial disc device, intended for the replacement of spinal discs in the neck that are damaged from cervical degenerative disk disease, had a 2-year human trial prior to FDA approval (Dooren 2006b). The manufacturer is planning a 10-year clinical trial follow-up or postapproval study to determine its effectiveness and safety in the long term. An additional postmarket animal study was recommended. The device apparently was considered no better than the spinal fusion treatment. In summary, long-term postapproval studies may be needed for medical devices.

Postmarket safety has become a major issue because of repeated media stories of device failures while in use in the human body and Congressional pressures for a more responsive FDA bureaucracy. Unfortunately, the FDA has relied on adverse event reports that are inadequate and on some reports from hospitals. These have been inappropriately reviewed and little action taken. There have been recent statements that the FDA will increase its public health response, which includes media outreach, urgent alerts, and enforcement actions. Thus, there is an uncertainty or a risk in relying on government monitoring of postmarket device safety, and that burden, in some form, must be assumed by others.

Biofilms

Medical devices that are implanted may become covered with biofilms composed of microbes. The fungus *Candida albicans* grows on surfaces first as cells, then develops in the form of an extensive extracellular matrix of carbohydrates and protein (Chin and Yeston 2006). There are studies under way to determine how *Candida* biofilms may be controlled. Antifungals apparently have limited value. Materials such as polyethylene for catheters may have a negative effect on biofilm development and adhesion. However, at the present time biofilms remain a risk for implanted devices since they may be a first step in other biological complications.

Biofilms are preferential protective habitats for hundreds of species of bacteria, including the persistent infections associated with medical implants (Everts 2006a). Staphylococcus, commonly found in hospital infections, may inhabit biofilms that keep antibiotics from diffusing inside. There may be group innate immunity responses that release antimicrobial peptides. This may help *Staphylococcus aureus* bacteria enter a host, express proteins that help with external colonization and attachment, and then initiate a general virulent infection process that may be deadly. Understanding this process led to the search for agents that could block the quorum-sensing communications to discern the pathogenic bacteria rather than creating an evolutionary imperative for drug resistance by killing the bacteria (Everts 2006a).

Blocking the chemical communications between bacteria may be helpful in reducing group virulence and the adverse effects of medical errors in infection control. This approach is somewhat similar to drug development based on ribonucleic acid interference (RNAi), which seeks to silence some genes. The RNAi therapeutic challenge is to target certain parts of the pathogen's cell for destruction, without harming surrounding tissue.

The medical problems and errors commonly associated with drug-resistant bacteria should not occur after there are pharmaceuticals developed based on the principles of chemical communications, RNAi, and tailored new vaccines that prevent bacterial adaptation.

Reprocessing

After initial use on a patient, some medical devices are reprocessed for further use on another patient. This includes devices such as trocars, laparoscopic graspers, and orthopedic pedicle screws. This may be done in a hospital, by an independent third-party reprocessor, by the original equipment manufacturer (OEM), or an unregistered (FDA) subcontractor. This reprocessing converts a single-use device to a reusable device. There have been some unfavorable media accounts about the reprocessing industry, so there may be some added risk in using such products. Some states now require the informed consent of a patient before the use of a reprocessed device that pierces the skin or enters the bloodstream.

The FDA requires a detachable label on the reprocessed device or its packaging, which eventually will be placed on the medical records of the patient; the label conspicuously displays the name of the reprocessor. Each device should indicate how many times it has been reprocessed and by whom (a history disclosure). Although reprocessors deny any added risk and may claim that a reprocessed device is substantially equivalent to a brand new device, there are those, particularly the OEM, who may insist there is some added risk. There may be undiscovered age- or use-related weaknesses in the reprocessed medical device. Medical devices may not be able to withstand repeated cleaning and sterilization or may be difficult to clean, and reprocessing may result in some loss of structural integrity.

Subcontracting

Medical device manufacturers may be large or small, but most delegate some function to subcontractors or vendors. This form of outsourcing may provide additional design skills, manufacturing capability, cost reduction for commonly used parts, specialized testing, unique precision tooling, and consulting services. There are risks in such outsourcing. For example, a commonly used component with some defects may adversely affect the products of several manufacturers. There is an analogy to the personal computer and laptop industry, which experienced massive recalls by several manufacturers because of a communality in defective batteries supplied by a vendor that subcontracted the battery manufacture. Just five well-known consumer electronics companies recalled nearly 7,000,000 laptops that used the same subcontracted part in their lithium ion battery packs. The section on reduction and integration in chapter 6 provides more information.

Portability

There are strong incentives for designs that result in portable or easily transported medical devices. The design objectives may be the miniaturization of components, compact assemblies of independent functions, reduction of costs, and consideration of the needs of underdeveloped markets. For example, a portable magnetic resonance imaging (MRI) system has been developed that requires no high-field magnets or cryogenics, has stage separation and optimization (for polarization of nuclear spins, spatial encoding, and detection), and has inexpensive components (Arnaud 2006a). Although there are opinions that such devices will never compete with other imaging systems, such as conventional MRI systems, the advances of modern technology can be surprising. New medical devices, using new technology, carry with them what is known as experimental device error or uncertainty risks. Portability also introduces a new set of risks.

Toxic Materials

A variety of materials is used in medical devices. Sometimes, a material is deemed toxic or could result in a toxic reaction. Polyvinyl chloride (PVC), a chlorinated plastic, forms dioxin during its production and when burned. The FDA issued a health risk advisory, but 4 years later it was still used in medical devices (Fontanazza 2006) despite its toxicity and possible carcinogenicity. In addition, di(2-ethylhexyl phthalate) (DEHP), a plasticizer used to soften PVC, can migrate out of the plastic from intravenous bags and tubing and into bodily fluids. DEHP is a reproductive toxicant. Alternative materials are available, such as nylon, Teflon, or silicone. When materials that could be determined in the future to be toxic are used, there is a risk in terms of customer satisfaction, litigation, and disposal.

Testing

The testing of medical devices is of special importance since there are limited opportunities to gather realistic human-oriented data before such devices are approved and implanted in humans. This suggests that there are uncertainties, unknowns, and unique risks during the early years of the marketing and use of the devices, perhaps until some of the implants reach the end of their life in a patient. That validation or proof-of-concept may last 10 years for some devices. This casts a special burden on user-surgeons and consumers-patients.

Most medical device manufacturers believe that they perform adequate testing or have alternative data supporting the presumed efficacy and safety of their devices. They are creative in terms of new devices, the potential financial returns are substantial, excessive delays are intolerable in a highly competitive marketplace, and they seem ready to take entrepreneurial risks. They feel protected by the FDA approvals and compliance with some of the key medical device standards. If this were sufficient, then why were approxi-

Table 4.1* Defibrillator Recalls (1996 to 2005)

Automated external defibrillators	
Distributed in 1996	~ 20,000
Distributed in 2005	~ 200,000
Malfunctions confirmed during attempted resuscitation of cardiac arrests[†]	370
FDA advisories (safety alerts) on the defibrillators or their accessories	~ 385,922

[*] Data from FDA Circulatory System Medical Advisory reports and 27th Annual Scientific Session of the Heart Rhythm Society.
[†] The most common causes of malfunction were electrical and software problems.

mately 20% of the automated external defibrillators distributed between 1996 and 2005 recalled?

The defibrillators are exemplars of devices that can malfunction, that need a system to locate and fix defective devices, and that are justified by a small number of malfunctions compared to the lives saved by the devices (a risk-benefit analysis). Recalls are an important consideration for all medical devices.

Testing is accomplished at each stage of the development of a medical device. At the conceptual stage, there may be evaluations by engineering and medical specialists, some preliminary tests, some computer simulation, and an examination of the performance of related devices from various sources, including the FDA MAUDE (Manufacturer and User Facility Device Experience) database. During the design stage, there may be benchtop testing, engineering analyses, specialized scientific research, and testing regarding the product's intended use applications. During the development process, there may be tests specified by harmonized trade standards (such as the International Electrotechnical Commission [IEC] 60601 safety standard), more lifelike test scenarios, accessory testing, and implantation of animals with the experimental devices. *Verification* tests determine whether the device meets company specifications. It is only at the stage of human clinical trials that a realistic evaluation can be made of the devices. Operator's manuals may be validated, particularly since the user may not have time to read the manual or uses it only when confused. *Validation* means that the device meets the user needs.

There may be failure analysis and testing of retrieved implant devices at any time. Among the conditions to be observed are wear, cracks, creepage, degradation, debris, deformation, looseness, clearance, migration, stress distributions, localized damage, erosion, electromagnetic interference (EMI) exposure feed-throughs, and fracture patterns. The harsh conditions of the human body may require alternative materials, including plastics such as polyethylene, different metal alloys, or ceramic coatings, as more compatible and less-bioactive substances. The goal may be to reduce cell reactions, particularly if there could be an active lifestyle and the patient's life span may exceed that of the implant. Hydroxyapatite coatings may be added to facilitate the bonding of implants to bone. It is expected that there will be constant improvements during the development and use of an implant. Thus, there are residual risks at each stage that are gradually eliminated.

The home-use defibrillator is a prescription-less, portable automated defibrillator that is low cost and simple to operate. These are used to intervene in sudden cardiac arrest, which claims 340,000 lives per year in the United States. They are operated by laypersons and may be found at airports, stadiums, shopping malls, and workplaces as public access defibrillators. They generally have thermal sensors and overload protection if lithium ion battery packs are used. They provide time, before calling emergency medical services, by shocking the heart into action so that others can provide some stabilization. It has been estimated that 40,000 lives a year could be saved with such devices. The user interface is critical, such as pads on the correct chest location, detecting patient motion, requiring a sequence of commands, and indicating whether the patient needs a shock. A wide range of user behavior is foreseeable.

Portable devices that are designed to be used by anyone available, without specialized training or instruction, have high risks during use. Rigorous human factors testing can substantially reduce or remove such risks.

Error Reduction

Learning Process

It is interesting to recall the learning process involved in the early use of implantable orthopedic devices. Early hip replacements involved frustrating errors in terms of strong and lasting materials, the specialized training needs of surgeons, what could be told to the patient, and the refinement of implantation procedures. Continuous problem solving and gradual improvement occurred. Lessons were learned that could be applied to the design and use of other future medical devices. Currently, many people believe that hip, knee, and shoulder replacements are fairly routine, can be considered commonplace, and are successfully performed in great numbers. This normal error reduction process took considerable time, but it did not involve the complexities of modern electronics technology.

Time Delays

The time delays of normal error reduction are illustrated by the following example. Diabetes (elevated blood sugar levels) occurs rather frequently since about 6% of the population has the serious problem of an inability to properly control blood glucose levels. There were measurement errors in reading, by eye, the early home use colorimetric urine test strips that were used to diagnose hyperglycemia. Later, fingerstick blood samples were used to diagnose both hyperglycemia and hypoglycemia. With handheld meters, using somewhat larger blood samples, accuracy improved, but there were painful fingerstick-induced avoidance, fear, inconvenience, and timing errors. Gradually, less-painful fingersticks, less blood, and less time were required, but the user (patient) still had to collect, measure, and interpret the blood glucose levels. Errors could be further reduced as glucose measurements become noninvasive (transdermal), automatic as opposed to manual, and continuous (frequent sampling) rather than infrequently taken at a given time. Further error reduction is possible if the monitor assesses profiles, patterns, or trends; sounds alarms; and is connected to an implantable insulin pump. The error reduction involved flows from the simplification or removal of human intervention in the process.

Formal Requirements

Requirements for human error reduction are generally found to be rather ambiguous or generalized in most standards, directives, and contracts for medical devices. It is not surprising that attempted compliance may be cursory, misdirected, token, or cosmetic. This contrasts with sometimes-meticulous conformance to the FDA Quality Systems Regulations (QSR) and Good Manufacturing Practices (GMP). Human error is often omitted in actual practice when using quality systems, reliability, and system safety standards. Most analysts do not have an objective and specific standard of care to use for error or a clearly defined methodology that is focused on human error reduction and prevention.

Analytical Techniques

There are books now available on analytical human error procedures (Peters and Peters 2006b) and compatible supporting books on workplace macroergonomics (Hendrick and Kleiner 2002) and system safety for design analysis (Ericson 2005). These techniques for the reduction of human error reflect sophisticated tasks based on appropriate scientific research, relevant test results, in-depth field reconstructions, and cost-effective evaluations of remediation efforts.

Analytical error reduction involves a diligent search for foreseeable, predictable, historical, product-unique, current test, and in-field life-cycle errors. The evaluation includes an analytic determination of root cause, true causation, and risk assessment as a predicate for determining appropriate countermeasures. The methodological rigor should match the needs of greater technical sophistication and the error consequences of modern medical devices.

Automation

The use of robots in manufacturing and assembly may save time, reduce manpower requirements, provide flexibility, improve quality, lessen overall costs, and provide for consistent parts inspection. Robots in an automated system for producing intricate medical devices can reduce cycle times, device variations, and discoverable defects. Automation often is intended to eliminate error-prone human assembly tasks and incomplete test procedures. Mistakeproofing devices may be necessary to prevent incorrect or flawed parts from assembly. The objective is to reduce errors caused by human intervention in the manufacturing process.

External Requirements

Standards

There are numerous regulatory constraints and applicable standards, both domestic and foreign, that deal directly or indirectly with medical devices and diagnostics services (American Society for Testing and Materials 2006, see appendix of this book, pages 205–211). They provide generally accepted protocol, methodology, and helpful techniques but may be burdensome, duplicative, costly, and time consuming and result in excessive documentation. The problem for device manufacturers is how to select what is relevant to particular design objectives and to appropriately consolidate if possible. Is compliance necessary for trade area certification, country registration, and the exacting performance that is usually necessary? Is something extra needed for proprietary leadership and avoidance of preventable risk? Can useful long-term data be accumulated and used to speed up future projects? Will the selected techniques facilitate translational applications from laboratory and clinical results to concepts of diagnostic and treatment devices?

There may be excessive requirements, in standards and regulations, intended to ensure a high degree of quality and safety during the life cycle of a medical device. Duplicative program elements and nonrelevant tasks should be eliminated. Harmonized standards should be given priority. Reliance on and compliance with standards do not provide sufficient assurance of device safety and expected performance. Standards may not cover key aspects of a particular device's performance. There are special topics to be added, such as specific requirements for error prevention and control. Greater effort should be expended to consolidate existing standards and clarify national protective requirements that may impede world trade. Organizational flexibility is necessary to respond in a timely and appropriate manner to future challenges yet conform to both mandatory and voluntary standards.

Outsourcing

When dealing with sourced components, there may be five problem areas of concern: (1) how to ensure the supplier's compliance with selected external standards; (2) how to ensure conformance with the prime contractor's contractual (internal) requirements; (3) how to ensure that there is appropriate control and systematic reduction of manufacturing variances in key attributes, variables, and the process; (4) how to deal with the need for continuous product and system improvement; and (5) how to determine whether the supplier has implemented a formal, separate, and effective effort to prevent and control unwanted error that may have undetectable hidden consequences.

Disciplinary Emphasis

There are disciplinary areas that may need emphasis, such as human factors related to cultural differences in the world marketplace. For example, human factors specialists may conduct observational studies to identify risk factors that may be related to a device's acceptance, use, and care in various cultures. This can generate hypotheses that are testable in clinical trials, need assessments, and studies of risk sensitivity.

There are areas that may involve prolonged analytical effort over the complete product life cycle, such as reliability and system safety. There are external requirements and criteria regarding how such analyses should be conducted (Andrews and Moss 2002; Ericson 2005) There may be significant product improvements that result. The question is how to accommodate such endeavors so they are meaningful and lean during each stage of product development. Similarly, quality system requirements may need to be tailored to product development milestones, objectives, and cost-benefit criteria.

These assurance disciplines are analogous to the metallurgical considerations in determining which material should be used in a given situation (J. R. Davis 2003). The balance may be between high strength and resistance to fatigue, corrosion, oxidation, abrasion, friction, bacteria, electrochemical activity, and hydrogen embrittlement. There may be special needs, such as shape memory, superplasticity, and superelasticity. There may be special requirements for hydrophilicity, biocompatibility, hemocompatibility, some surface treatment, or coating. The choice may be some combination of materials or an alloy of nitinol (nickel-titanium), titanium, stainless steel, cobalt-chromium, ceramics, or plastics. The material choices are compromises, and they affect design alternatives and reliability strategies. Simply put, the external requirements dictate design.

Communications

Effectiveness

Communications among the medical device manufacturer, the physician, the patient, various government regulators, and others can become a serious problem area. In assessing communication effectiveness, questions arise about the desired transparency of useful information, its level of detail, its explicitness, and the protection of proprietary information. Is the information tailored appropriately, measured regarding education level, and valid for any presumptive use by trained and sophisticated intermediaries?

Complications

The messages may be complicated by differing perceptions of need, personal attitudes toward instructions and training content, and the levels of comprehension of the targeted audience. The means of communication may be oral, print, or electronic (particularly the Internet). It includes advisories and warnings (Peters and Peters 1999), exercised recall (Mundel 1999), and public information releases. There is usually a need for multiple instructional approaches and the continuous availability of the latest source of information. This includes physician manuals, patient handbooks, and messages about applicable research, clinical, and in-field findings. There may be training sessions, lectures at professional meetings, and customer visitations. There are reports to government agencies, labels on the product, descriptive material on the packaging, and package inserts.

Ambiguity

Communications should be carefully reviewed to avoid miscues, misunderstandings, omissions, misdirections, and undesirable inferences. Common sense is untrustworthy. Such matters are best left to those aware of the scientific, legal, behavioral, and risk mitigation aspects of human-to-human communications. It is important to determine the error tolerance limit.

Even with the FDA medical device reporting requirements (21 CFR Part 803), there has been professed ambiguity regarding what constitutes an adverse reportable event. If ambiguity can exist in such interpretations among sophisticated parties, then it may be reasonable to expect the unsophisticated users to have even greater ambiguous interpretations of communications directed to them.

Content

Are the risks of restenosis (artery reclogging) effectively and realistically communicated to the patient prior to the use of coronary drug-eluting stents? Are those communications tested and standardized, and do they constitute reasonably effective messages regarding the risks and benefits? The risks may differ for embolic protective filtration devices intended to capture dislodged material during cardiovascular intervention. Again, the risk communication is different for pacemakers and defibrillators, artificial discs, specific-site drug delivery devices, or osteotomy systems (including orthopedic fracture fixation plates, nails, screws, washers, wedges, and allografts). There may be estimations or measurements of depth, angle, force direction and magnitude, orientation, or pathways between tissues. There may be special communication problems in dealing with minimally invasive surgery and telemedicine. Patients fully informed tend to establish trust bonding to professionals and companies. This is much better than having them develop hostile attitudes based on misunderstandings, improper expectations, or other communication failures.

Product Design

Essentials

Product design is inherently an iterative, successive, step-by-step, trial-and-error process. Considerable time is consumed from concept to marketability. Thought should be given to refinement of the process by using a more efficient systems integration approach, doing some consolidation and overlap of specified functions where advisable, and utilizing multiple parallel pathways for different product development functions (e.g., earlier and more frequent formal design reviews for better coordination, guidance, assurance, and navigation between the choice and decisions encountered in the design review process). Human factors studies should start at concept design rather than late in the development stage. Early starts also include reliability and system safety. Condensation of design schedule time, while retaining high quality and management objectives, is forced on us by current events, such as emerging worldwide industry competition. Postmarket (after-sale) surveillance and rapid corrective actions have become critical. Sell-and-forget or consideration of surveillance as nonproductive create unnecessary avoidable problems. Surveillance should include early management attention to reports of nonconformities and complaints, quality indicator trends, the findings of direct audits, and surveys of compliance with customer expectations. A good surveillance program should reveal even marginal human errors in such detail that causation and remedy can be determined since it is a difficult transition from common beliefs about error causation to more valid explanations and effective countermeasures. The normal error reduction process is unacceptable given current societal expectations and product competitive needs. Appropriate management of risk reduction, when perceived as a nondelegable oversight responsibility, may provide proprietary advantages in a highly competitive world marketplace.

World Trade Problems

The world's population is increasing by 1 billion people every 11 years, suggesting a rapidly expanding marketplace for medical devices. The population growth is greater in urban areas, suggesting some concentration of marketing opportunities. The primary problem is affordability since vast economic inequities remain, although the gap is narrowing. The second problem is the proliferation of local protective standards since the vitality of world trade depends on harmonized or generally accepted product requirements.

Expanded world trade may create a number of human error prevention problems for medical devices. The patient or user population in various sections of the world may have differences in relevant cultural, common practice, habit, custom, individual capability of due care, and personal responsibility issues. Medical specialists have a varied knowledge base, mode of practice, training, beliefs, and value systems. Language differences may create communications problems, although English is gaining as a common language. There may be an emphasis on low cost in terms of initial price or price transparency, maintenance, service, repair, explant, replacement, and examinations by medical specialists. There may be problems concerning intellectual property rights and counterfeit parts. Each source of human behavioral variance increases the likelihood of human error in some respect. User-friendly design may be difficult without a realistic cultural understanding.

Customer Satisfaction

Throughout the world, there has been a general increase in access by the general population to information about products and services. This has been accompanied by greater expectations of customer satisfaction directed at individual needs. Customization, whether individual or group, is costly and error prone. However, an unfilled market creates competition. It is apparent the emerging markets may present novel opportunities that impinge on product and system performance, corporate financial performance, enterprise social relief performance, and error prevention performance. It is an important management decision regarding the assignment and function of those responsible for expertise in cultural differences since translated words may have unexpected meanings, and ethnic and national biases may abound; the expert should be in a position to deal directly with design, test, and error reduction specialists.

Critical Comments

An article in *Forbes* magazine (Herper and Langreth 2006) discussed the business of implantable medical devices as an $80 billion a year business with 20 million Americans implanted with "miracle gadgets" or "man made body parts," including

- 160,000 hip replacements per year
- 280,000 knee replacements per year
- 250,000 cardioverter defibrillators per year
- 1 million heart stents each year

The article indicated that there have been design flaws, malfunctions, and poor delivery systems (implant procedures). There are 4,000 new devices approved each year, but from 2002 to 2007, 79 were recalled for potential fatal side effects, and 2,300 were recalled for lesser complications. One company now faces 550 individual and class action lawsuits. The article talked about "acquiescent regulators, gadget-happy cardiologists and surgeons," and "pushy device makers" as responsible for the "device boom." Note that such critical comments in a leading business periodical should be taken seriously.

About the same time, the chief science and health correspondent for the NBC news network stated that "millions of Americans could be walking around with tiny time bombs in their hearts" (Bazell 2006). This refers to drug-coated stents used to treat coronary heart disease from arteries clogged with cholesterol-containing plaque. Originally, tiny balloons were inserted to push open the clogged arteries (angioplasty), but in many cases the artery would close again (restenosis). In 1994, cardiologists placed wire meshes (stents) around the balloons and left them in the artery to prop them open. Cells grew on the stents and clogged the artery, so drug-eluting stents were used to provide a drug that prevents cell growth. The cell proliferation was effectively stopped, but some patients suffered fatal heart attacks from what was claimed to be quick blood clot blockages that formed on the stents. There were estimates of 2,000 deaths per year. The stents cannot be removed easily, so anti-blood-clotting medications (such as Plavix) were used, each with their own risks. The NBC correspondent, Robert Bazell, stated that "doctors and scientists admit they are in unchartered waters with a frightening problem that was unanticipated" and "what should the estimated 4 million patients who already have the [drug-eluting stent] do?" Note, does this portend other medical device problems, or has the industry already undertaken effective preventive measures?

Conclusions

The unique vulnerabilities of the medical device industry necessitate a more vigorous, comprehensive, and well-organized risk reduction effort that results in effective controls and surveillance. The near future will involve more device complexity, design creativity, foreign competition, medical error (patient safety) constraints, and regulatory burdens. It is reasonable to expect less tolerance for adverse events and failures to meet customer expectations. There is particular vulnerability concerning preventable human errors because common emotionally tainted beliefs may override rational evidence-based explanations for specific countermeasures. The elements of a good error management program should be applied contemporarily to the design of future products. Appropriate advisories and upgrades may help to retain customer contact, promote satisfaction, and foster market acceptance for future products.

Caveats

Risks

Medical devices pose unique risks that necessitate special precautions in design, manufacture, application, and use. There exists a delicate balance between perceived risks and the social value of medical devices and therapeutic equipment.

Analysis

Special detailed analytic techniques should be utilized from concept design to ultimate disposal of the product. Analysis permits and ensures prevention. Without appropriate analysis, there may be unrecognized and unacceptable risks. Uncertainty can be costly.

Culture

In an age of world trade and population diffusion, there are cultural differences that need to be identified and countermeasures undertaken.

Corrective Action

The conventional post facto correction of medical error problems is unacceptable if there are risks of serious preventable injury. A specific program for medical error management action is advisable since appropriate analytic techniques, validation, protocols, and monitoring procedures are available for application at reasonable costs.

Medical Communication

The surgeon who implants a medical device is generally considered a learned intermediary, someone who has special knowledge about the medical device. That knowledge is determined by what the company elects to communicate in the form of printed manuals and the advice of company representatives. Marketing influence should not overcome the transparency needed to adequately communicate possible problems for the surgeon and the patient.

Recalls

A recall plan should be in place to update warnings and advisories, to initiate field changes or improvements to the products, to respond appropriately to discernible trends of malfunctions and explanations, to recall or replace components beyond wear-out or approaching service life expiration, and those manifesting a serious defect. The plans should be exercised or rehearsed for quick reactions at reasonable costs in a manner that does not damage public relations and brand images.

Biofilms

Bacterial communities grow on human tissues, medical devices, or some objects that come into contact with human tissues. A catheter may have a chemical communication network regulating the growth of a biofilm by quorum sensing. Such biofilms may cause up to two-thirds of all serious infections ("Mass Spec Analysis" 2007). Appropriate infection control measures should be built into medical devices.

Ease of Use

Medical devices should have a human factors analysis that has a focus on error-free ease of use. An intuitive design is generally preferred because people are often reluctant to read user manuals or follow complex procedures with sufficient due diligence. Symbols, icons, and graphics in foreign countries may or may not overcome basic language problems. Cultural differences in use patterns or customary practices should be determined, at the on-site location for context, using representative clusters of hospitals, geographic regions, and countries. Preliminary information, for planning purposes, may be obtained by surveys, telephone calls, or e-mails. The final product should be considered a system that includes the hardware, software, caregivers, and patient; all are interacting according to local custom.

Life Testing

Highly accelerated life testing, whether performed on devices or humans, may be performed on small-size samples. The objective is to identify errors and weak points in design at an early stage. Such testing may not reproduce failure modes at a precise stress level, as opposed to either a stress range or an increasing stress beyond the specification limits. Well-written test reports permit validation and repeatability of the testing and can assist in identifying appropriate corrective action before normal marketing reveals what might be called delayed unanticipated discrepancies or defects. Varied stress levels may provide data for risk assessments that contain both worst-case and central (expected) estimates. Life testing should include a broad range of incorrect and unexpected user actions or inputs in an error simulation approach.

Degree of Safety

The European Commission presently utilizes the concept of *reasonable practicality* in assessing whether there is an acceptable degree of safety. This is a combination of the foreseeability of harm and the technical feasibility of reducing or eliminating the risks. They are considering more precise definitions that increase the level of safety while providing some developmental exclusions. In an era of worldwide business enterprises, each state, country, or trade region may have precise definitions of legal fault or blame, but all are derived from old common law principles and recent court interpretations that balance the severity of anticipated harm against prudent precautions to avoid the harm. The degree of safety is always somewhat ambiguous and changing; it is not a fine line for easily defined compliance. Medical devices, regardless of marketing limitations or restrictions, may enter the jurisdiction of almost any country in the world.

Open Communications

The contract supplier may be reluctant to disclose to the OEM information that is considered proprietary or a trade secret, an intellectual property right, or even a manufacturing method that includes quality assurance. The OEM may be equally reluctant to discuss detailed product specifications, raw material requirements, marketing forecasts, and other information that is needed by the supplier to calculate manufacturing cost estimates. Each side is effectively blindfolded and must employ generalizations, and this invites error. More realistic discussions may result from the use of nondisclosure agreements. Some countries have policies that foster open communications so that all parties benefit from cross fertilization.

Remote Patient Management

An implantable medical device may have a remote monitoring system. A wireless patient management device could check the implanted device on a regular schedule; a receiver could collect the data and determine undesirable data excursions and, when necessary, could send an alert or compressed data to a remote medical professional at a secure Web site. Life-threatening deviations in bodily functions can be detected and treated at an early stage. In the future, such medical devices could extend hospital-type services to at-home patients. As such advances in technology are transferred to home locations, it may invite correctable error, as well as shift supervision, reduced costs, and improved patient safety.

5

Analysis

The investigation of medical errors can be relatively informal and haphazard, performed by those who are immediately available and well intentioned, who apply what seems to be good common sense under the circumstances and with just a vague hope that the problems will then go away. In contrast, there could be a dedicated analysis performed in an organized manner by a specialist trained in such matters. Fortunately, there are proven methodologies, with refined techniques, that can be applied by highly experienced specialists in such fields. This includes adaptations of system safety, human factors, reliability engineering, quality assurance, industrial engineering, and human error analysis. This chapter describes some of the analytic approaches that could be applied to the control of medical error and indicating the benefits and limitations of each approach. A choice of techniques can be made based on need but much depends on the individuals assigned to such efforts and their personal talents and skills. The analysis guidelines and procedures that are presented could vastly improve many current disorganized efforts at error reduction.

Corrective Action

If a meaningful error occurred, then the traditional approach was to investigate that narrow situation or set of circumstances and to focus on finding a probable cause and an acceptable corrective action. The techniques used were a variant of accident investigation and reconstruction, the use of a logic-based analysis such as in troubleshooting, and if repetition of the error occurred, digging deeper into causation using root cause techniques or repeated why-who questioning in a search for more remote or hidden causes that could be subject to effective corrective action (Kepner and Tregoe 1965). This was and is a *post facto* analysis of a single event or type of error. Essentially, it is a wait-and-see form of trial and error. There is uncertainty regarding when, how, where, and what kind of error may become manifest. It may be based on the hope that the status quo is acceptable or desirable, and one error is just a temporary deviation that should be corrected for a return to normalcy.

Preventive Action

The advent of complex systems and processes such as nuclear weapons, missiles, space, and electronic systems created situations in which a single error, mistake, or failure could lead to catastrophic results. Correction of errors would be too late since the damage would have already occurred. Prevention became paramount. Various analytic techniques were formulated to identify possible problems before they could become manifest. Reliability engineering specialists used what is called failure mode, effects, and criticality analysis during the early design and development stages to identify, categorize, estimate frequency and severity, and propose remedies for possible equipment failures. This technique is utilized to weed out possible failures in medical device hardware and to overcome design weaknesses in medical device software, but it generally omits consideration of human error and human factors.

System Analysis

A broader perspective than just a focus on hardware components and system design is found in system safety analysis (U.S. Department of Defense 2000; Ericson 2005). This includes what is called *fault tree analysis*. The intent is to remove uncertainties in the use, life cycle, reconfiguration, and ultimate disposal of a system. It is an early, detailed, and preventive approach to design in general. It is characterized by three recognizable functions: (1) a standardized protocol or specification for universal use that can be required by contract or management direction; (2) an organized attempt to focus on all parts of a complete system, including machine and human interactions; and (3) the use of more objective data for risk assessment and risk reduction, including interdisciplinary design coordination and oversight activities.

Human Error Control

An analytic approach used in system safety or applied as an independent focus on error prevention is the technique of human error management or control (as detailed in Peters and Peters 2006b). This approach covers all sources of error in a complete system by the detailed identification of all possible errors capable of causing harm, assessing their delectability and correctability, evaluating their risk on an expected frequency and severity basis, and determining the effect of various controls that could be reasonable, cost effective, and used without creating appreciable side effects. It includes consideration of human behavioral vectors, including personality traits and disorders and the reactions, effects, and control of unwanted variance from such predictable human behavior (Peters and Peters 2006a). The techniques are applicable to the design of medical devices, hospital equipment, and hospital workplaces.

Risk Assessment

The term *risk assessment* has many different meanings around the world. The harmonized international standard on the Application of Risk Management to Medical Devices (ISO/FDIS 14971:2005) indicates that *risk analysis* uses available information, and *risk estimation* is a conclusion based on the likelihood of the event and its severity of harm. In essence, if there is an error about which a risk analysis can be performed, then an assessment or estimation of the amount of risk can be opined, and it can then be categorized as an acceptable or unacceptable risk. If the risk can be reduced by some means (see chapters 6 and 7 in Peters and Peters 2006b), a *residual risk* is determined. Warnings should be used only to further reduce the remaining or residual risk.

Risk assessment can serve as an indicator of the status quo, benchmark, background, or *predicted risk*. The implementation of various remedies results in *modified risks* at various costs and effects. The final residual risk is the *manifestable risk*. The risk considered acceptable is the *tolerable risk*. Thus, an error analyst may indicate, for each predicted error, that it is minor (acceptable), moderate (action required), or major (intolerable). The objective of the analyst is to use risk mitigation techniques to achieve ALARP (as low as reasonably possible) residual risks. In medical research situations, the term *relative risk* (the excess over background risk) may be used if appropriate.

Other analysts may use other risk assessment procedures, such as risk-benefit analysis, cost-benefit analysis, risk-utility analysis, or a reasonably practical analysis. There may be some unavoidable risks outweighed by emergency circumstances, financial interests, or political factors. Some international standards, for example, those from the British Standards Institute (such as BSI-OHSAS 18001:1999 and BSI-OHSAS 18002:2002) and those from the International Labor Organization (such as ILO-OSH:2001), state that risk assessment is simply the overall process of estimating risk and determining whether it is a tolerable risk.

In studies of hospital workplaces, the dominant failures were in risk assessment and risk control. The risk assessments had not been implemented robustly (Peters and Peters 2006b, p. 28). If a risk assessment is not performed, then it is generally assumed that not enough is known about the error, its cause, and its correction.

In the evaluation of drugs and devices, the total increased number of quality-adjusted life years may be compared with the overall cost or expense. A cost above $50,000 a year may mean a procedure, drug, or device is not warranted on a cost-benefit basis ("Baby Talk" 2006). The use of a fixed cost in risk evaluations has been criticized for many years by experts who ask how much is a life actually worth to the person, to the parents, in terms of infection control, and considering the possible consequential costs of a disability.

Mistaken Beliefs

It is generally recognized that sloganeering is ineffective. This includes "be careful" and "avoid mistakes." Simplistic slogans are ignored, quickly forgotten, and so general they are ineffective.

Similarly, there is a widespread belief that "foolproofing" and "goofproofing" are impossible. This may serve to excuse a failure to do what is actually possible and needed.

In terms of slogans or excuses, this reveals a lack of sophisticated knowledge regarding error proofing or a lack of motivation to utilize existing techniques for the reduction of medical error. Such techniques should be logical, evidence based, realistic, and multidisciplinary in character so they are productive, reliable, and useful.

Observational Demeanor

Information concerning errors should be gathered by unobtrusive observations since those who know they are under observation will modify their behavior to conform with what they believe to be correct or expected by others.

Similarly, in interviews or conference sessions, become a professional listener. Do not coach, recommend themes (other than problem solving), suggest answers, provide concepts, or indicate desired guidelines. Do not infer accountability, personal responsibility, or penalties. Attempt to achieve mutual confidence, neutrality, a circumspect demeanor, honesty, trusting relationship, and rapport. Focus on unlocking perceived shortcomings and self-protective behavior.

Be innovative in achieving *direct* access to the information needed for error analysis and prevention.

Correct Terminology

Learn the terminology of those performing error analysis tasks to ensure accuracy and precision in understanding and fully comprehending the words used. For example, *remedies* suggest complete solutions, whereas *countermeasures* suggest less-effective means to control error. The cues, prompts, coding, or signals relied on may be appropriate, or they may be *false cues* that mislead and result in error. There may be *clutter* or unnecessary information that can slow, confuse, or distract from the desired human performance, such as informed and correct decision making. *Task analysis* is a detailed and sequenced set of behaviors that may be correct in terms of desired procedures or may contain violations of what is considered prescribed or desired behavior. *Focus groups* include a small number of people, apparently independent, selected according to some type of demographics or user population and used during the analytic process to provide spontaneous reactions related to preferences, habits, values, and desired attributes about the subject of the intensive focus. *Consumer clinics* are often a form of a focus group. There are many more specialized terms defined throughout this book.

Complete Process

The error analysis should include something more than usability, operability, human use in the intended function, prescribed procedures, or sequence of events surrounding the medical error. In terms of medical devices, instruments, and equipment, there may be consideration of selection according to need, transportation and receipt, inspection and calibration, installation and break-in, type of prior use, servicing and repair, recycling, and inerting for forensic examination or ultimate disposal. A component damaged in transit may manifest a time-delayed failure. Post-error changes are always important. The investigation may be an iterative step-by-step process that attempts to reveal weakness in the complete process rather than a quick fix of just one element of the process.

Post-Control Measures

The error analysis is not completed on the implementation of recommended countermeasures and remedies. There should be immediate monitoring to confirm and ensure that the unwanted behavior has changed in a positive manner. There should be periodic continued surveillance to ensure that there is no regression or modification of the error control measures.

Some trade standards recommend a documental risk assessment plan that is required to be updated and placed in a risk management file. This includes how compliance was achieved for applicable trade or professional standards. It may include entries for monitoring and surveillance of risk control measures under the stresses, demands, and conditions of normal use, performance, or compliance with procedures. In essence, what is the proof that the error control procedure or modification actually works in real-life situations?

Operational Discipline

A frequent error category is a lack of *operational discipline*, that is, a failure to follow the required, customary, or desired procedures and stay within the operational envelope defined by the procedures. The error analysis should consider whether the procedures are oral or written, whether they are complete for each step or task, whether they are too difficult to comprehend and follow, whether they are commensurate with the educational level and literacy of the user, and whether procedures are consulted only when steps are forgotten or problems arise. Are written procedures conveniently available, practical, and understandable? The analysis should focus on exactly why the procedures were not followed and what is required for operational discipline in terms of training, instruction, supervision, or management.

Error Troubleshooting

There may be first impressions that a medical error has occurred or was a contributing factor in the causation of an undesirable event. The impressions may suggest mistakes, wrongful conduct, personal fault, improper behavior, or violations of procedures, rules, or regulations. Such impressions as trouble reports should be investigated as quickly as possible since relevant facts tend to change, fade, and become forgotten. Too much elapsed time may permit the same conditions and error to reoccur unexpectedly. Telephone interviews, reviews of available incident records, and evaluations of prior histories or experience may help to form a preliminary causation scenario. The collected information could be helpful to a professional-level investigation conducted subsequently. Such preliminary reports often contain biased subjective opinions and ostensible evidence that may be incorrect, but they generally contain causation hypotheses to be tested, factual information to be verified, and likely corrective actions.

Traceability

Modern electronic computer systems provide a means by which the source of many errors can be quickly determined. For example, they could provide a track-and-trace historical record of drug prescriptions, medication identification and interactions, the match of drug and patient identification numbers, how and when the drug was distributed, the dosage levels consumed, the patient reactions, and subsequent treatment. There may be patient bar code bracelets, smart medicine-dispensing carts, smart drug labels, and other checkpoints for the computer system. Such computer system traceability may be applied to other processes that could be prone to error or mistakes. The computer system could gradually accumulate information (data) on the frequency and outcome severity of various categories of error. It could sound alarms for impending error or near misses. Traceability could be an important first step in gathering relevant data for a sophisticated error analysis effort.

Industrial Engineering

Industrial engineers attempt to improve process control and monitor process variations. The hospital patient flow could be considered a process that needs control, requires continued improvement for unit cost reductions, and needs to be monitored for productivity and error control. Industrial engineers are specialists in methods improvement, production control, and quality control and are involved in a wide scope of activities related to industrial manufacturing. They are familiar with motivational concepts such as that of the French philosopher Rousseau, who in 1762 indicated that organizations flourish only when both individual and collective interests can be pursued to the fullest extent possible. They are familiar with the more recent concepts of Maslow, McGregor, and the Hawthorne study and their application to current sociotechnical complexities. There are academic industrial engineering programs that award the doctoral of philosophy degree, some industrial engineers have benefited from multidisciplinary engineering programs conducted at some universities, and many minor in human factors engineering. Among the industrial engineering concepts that are important are that people are motivated by (1) work made interesting, (2) the opportunity to contribute to decision making, and (3) personal recognition of good job performance. They apply such techniques and concepts to a wide variety of applications. Only a few may be found in medical settings, but they possess valuable quantitative analytical skills.

Quality Assurance

Introduction

The word *quality* has been broadly used in relation to medical devices and hospital audits. Quality concepts have been widely discussed as they pertain to error elimination since quality control was founded on the inspection of parts to ensure component uniformity and error-free compliance with design specifications. Quality specialists may be found on the staff of medical device manufacturers, and they are generally involved with inspection, testing, and warranty activities. They function to ensure the quality of the product for customer satisfaction and to ensure control over process variation. The following examples illustrate the work accomplished in reducing medical error by quality specialists and industrial engineers.

Drug Delivery

The application of quality techniques to health care problems has had some reported success. The use of a Six Sigma approach to analyze a hospital's medication use process (the patient drug delivery system) resulted in streamlining the process ("Six Sigma" 2006). There were 42 steps eliminated from 132 steps in the process. The time to dispense and administer the drugs was reduced to 104 minutes from 186 minutes. There was a claimed reduction of 70% in medication errors. In such situations, there is almost a religious zeal in applying what is known as lean Six Sigma techniques.

Change Management

At another hospital system, there were six waves of Six Sigma on 60 projects. This included training classes on "change management" or change acceleration by overcoming cultural barriers, creating shared needs, and mobilizing commitment. There were also training classes on "fast track decision making" by team involvement, in-meeting decisions, and empowering people closest to the process. The training of hospital personnel involved analysis of the value of fixing deficiencies and isolating major areas of improvement while assessing the costs of change.

Bed Assignment

One hospital used Six Sigma to reduce delays in bed assignment turnaround time, including postanesthesia care and the emergency department (Pellicone and Martocci 2006). The focus was on patient flows that led to delays in operating room throughputs and emergency department holds. They developed a process map and a cause-and-effect diagram for all variables, calculated defects per million opportunities, and applied statistical analysis to determine significant differences in the process by shift and time needed.

Medication Administration

A midsize hospital project used Six Sigma to reduce medication errors (Esimai 2005). The medication administration involved six steps: (1) selecting and procuring, (2) storing, (3) ordering and transcribing, (4) preparing and dispensing, (5) administering the medication, and (6) monitoring medication effects. The project concentrated on order entry and reducing medication administration record (MAR) errors. It reviewed and verified process maps. It established an overall MAR process error rate of 3,300 per million opportunities or 0.33%. This was reduced to 0.14% in 5 months. The annual cost reduction was $1.32 million. The errors included the following: compared to the original faxed medication order, there were additional instructions not inputed, wrong doses, and wrong drugs; duplicate order entry with two different prescription numbers; medication frequency errors; medication omissions; orders not discontinued; orders not received; orders for wrong patient; and incorrect route (intravenous or intramuscular). High error frequencies resulted from misunderstandings of guidelines and instructions by pharmacists, which were correctable by remedial education and closer supervision.

Pharmacy Errors

Another study attempted to reduce medication error in a hospital setting. The error was defined as information copied incorrectly or omitted (a transcription error), such as missing, inaccurate, or partially provided special instructions with a physician order (Benetez 2007). The cause of some mistakes was disruption to the pharmacist during the order entry process. To reduce disruptions, all intravenous fluid orders were to be sent to the pharmacy before 6:00 a.m. for early preparation. Missing medication sheets for clinical units were transmitted by fax rather than phoning questions. Illegibility of the physician's order was improved by more room on the medication order. The error reduction goals were achieved. In the process of the study, they identified nurses' needs as quick access to a patient's medication information, quick pharmacy turnaround, a history of patient medications, portable and mobile information, double-checking capability, and trustworthy order entry.

Quality Techniques

Total Quality Management (TQM) generally takes a process that is out of control and applies quality techniques to get it back in control. Six Sigma takes a process that is in control and applies quality techniques to reduce variation within the process and improve performance indicators. The Six Sigma procedural stages are to define the project and goals, measure the process as currently configured, analyze the data, improve the process in terms of defects, and control the process to maintain the improvement. The objective of a Six Sigma approach is no more than 3.4 defects per million opportunities. The Lean concepts include having available only the things that are needed, eliminating wasted motion, providing housekeeping to identify equipment problems, documenting to ensure that all persons do their jobs in the same manner, and using checklists to make sure controls stay in place. The fundamental question is whether a manufacturing process model can be applied to the different business models found in health care organizations.

The quality assurance successes in health care systems may be a function of just plain diligent analytic efforts by persons who customarily inspect, audit, and foster improvements in operational processes, equipment, and products. The Six Sigma criterion or objective of processes that create no more than 3.4 defects per million opportunities is achieved by "the relentless and rigorous pursuit of the reduction in variance of all critical processes in an organization" (J. A. Johnson et al. 2006). Design of experiments (DOE) has been used with the Six Sigma approach. This consists of manipulating variables in an organized way to determine what could be changed to improve the process (Conklin 2004). It can be used with an analysis of variance (ANOVA) statistical analysis to determine if a change in an independent variable changes the response variable. It helps to determine interactions (dependence) between the independent variables.

Quality Standards

There have been a number of international standards pertaining to quality. Based on the 1994 International Organization for Standardization (ISO) 9001 quality management standard, an ISO standard for Quality Systems—Medical Devices (ISO 13485) was issued in 1996 and revised in 2003. Medical device manufacturers are registered for and assessed regarding their compliance with that standard (ISO 13485:2003). There is also a guidance document (ISO/TR 14969) that provides guidance for the intended use and implementation of the quality systems standard. The standard includes a subclause (8.2.2) that requires internal audits of the quality management system (subclause 1.1). There also are risk management procedures and a requirement for a quality manual. The European Union has a Medical Device Directive, Canada has Medical Device Regulations, and Japan has a Pharmaceutical Affairs Law.

The widespread adoption and compliance with the quality systems standard ISO 9001 is suggested by the fact that there were 776,608 organizations registered to the standard in 161 countries as of December 2005. There are quality specialists who are certified to perform the functions of a lead auditor, internal quality systems auditor, or internal quality systems auditor relating to the implementation of the ISO 9001:2000 quality system standard. Such specialists also may be certified as an environmental management systems internal auditor or lead auditor relating to the ISO 14001:2004 environmental standards.

The attempt to apply automotive industry process methodology to health care processes involved the use of modified ISO 9000 quality standards, such as QS-9000, which is utilized in the automotive industry. This application of engineering principles to medical services resulted in derivative standards, such as HC1 (ISO IWA-1) and BOS 2008. The idea was to identify common chronic conditions in the delivery of health services and, if routine and predictable, standardize them. It was found that the language of the standards was not easily translated into "healthcare speak" (Reid 2006). It was concluded that "healthcare practitioners can embrace a push for excellence" and "this should improve patient safety over time." One of the objectives, in ISO IWA-1, is called *error proofing*. Other objectives are broad in scope and wide ranging, including the adoption of disease management programs, cost monitoring, internal auditing, and management system auditing.

Limitations

Despite the emphasis on quality control (focused on inspections) and quality assurance (a broader term), there have been many medical devices, used by users or customer-patients, that have malfunctioned and required replacement or recall. This suggests that the quality techniques, intended to prevent devices that might fail or malfunction from reaching the marketplace, need some improvement or more widespread application. To some degree, the quality concepts intended to provide worker motivation and product uniformity may have been misinterpreted. For example, one certified quality manager at a medical device manufacturer indicated that quality begins with the production worker, and no further inspection is really necessary, that is, quality is free, do it right the first time. Others use only a sampling process during the only inspection, which is at the final assembly. Another interpretation was observed in a different industry in which the assembled products in one factory were given a short functional test at the end of final assembly; if the product failed the test, then an abbreviated inspection regarding cause occurred. In other words, quality inspection requirements may be inconsistently applied depending on the interpretation of the writings of some quality gurus, such as W. Edwards Deming, Joseph M. Juran, Genichi Taguchi, Gemba Kaizen, Philip Crosby, and Frank Gryna.

Quality Programs

Among the quality philosophies, or programs, with advocates and critics for each are Total Quality Management, Quality Is Free, Zero Defects, Quality Circles, the Taguchi methods, the Six Sigma Strategy, and other inspirational approaches to achieving high levels of quality. They motivate quality professionals to apply certain promising techniques to achieve better results than other customary approaches.

The fundamental questions are: What philosophical constructs motivate the performance of a particular quality specialist, and how would this apply to specific medical error problems? Could the quality specialist work harmoniously and not disruptively with those from other disciplines?

Such efforts seem to have merit in performing novel administrative audit functions relating to patient safety.

Caution

A recent report in the general membership publication ("Too Many" 2006) of the American Society for Quality (formerly the American Society for Quality Control) indicated that hospital quality reporting systems varied widely in many respects. A study conducted by the Center for Studying Health Systems found that the reporting requirements were a burden to staff, that there was little evidence that hospital reporting systems actually improve the quality of care, that the difficulty in fulfilling the complicated requirements might not be worth it, and that there should be increased coordination between programs.

Other Disciplines

There are other disciplines that could provide analyses, including those of registered nurses, pharmacists, and technical representatives of medical device manufacturers. Chemical engineers who specialize in process safety management may have experience in finding hazards and determining risks at chemical plants. They identify, classify, and evaluate process hazards as a preventive measure. This often includes consideration of human factors, loss prevention, and incident causation. They may be able to apply this specialized knowledge to medical error problems.

There are those have specialized in reliability engineering or reliability management who perform analyses of past failures and predict possible future failures. They may be able to reduce risks, by design improvements, of or from medical devices and equipment.

Government and Industry Reviews

There may be government publications, domestic or foreign, that focus on a particular medical error and what could be done about it. Separate from regulations, there are "educational documents" for "guidance purposes" that "do not establish legally enforceable responsibilities" for medical device manufacturers and distributors, clinicians, health care workers, health care facilities, patients and their families, and others who may have an interest. Similarly, there are publications of the state hospital associations, a robust body of literature found in medically oriented journals, funded research reports, litigation reports, and many industry, trade, and company publications. An Internet search is a preliminary part of any analysis of medical error so that there is a benefit from the experiences of others. Such publications should be viewed critically, with emphasis on objective methodology and the degree that conclusions are relevant, implementable, and cost-effective.

In-House Teams

It may be appropriate to train members of a medical error improvement team who are staff members of the hospital or health care organization. They can accumulate special error reduction knowledge, maintain confidentiality, and not disrupt the culture and essential function of the entity. This is a self-help process and should be led by a physician, nurse, clinical engineer, or professional with extensive clinical or specialized experience at that facility. There may be problems if the team is perceived as investigating themselves, as staff evaluating staff peers.

The need for clinical experience is suggested by the following scenario in which some may claim medical error, and others may claim no error. A Papanicolaou test (Pap smear) may contain endocervical cells diagnostic of low-grade endocervical adenocarcinoma. Is it medical error to diagnose cancer? It may be reactive endocervical cells found in benign Pap smears (Foucar 2006).

The team leader may be selected or rotated according to the type of error manifested. If there are many different types of error, then the Pareto principle (worst first) may suggest first developing lists that rank each error by the magnitude of risk or by the ease of remedy. It is important that the process involve detailed analyses in the attempt to uncover true causes rather than superficial quick fixes that may prove ineffective in the long term.

Personality Factors

Any analysis is dependent on the attitudinal, motivational, and philosophical state of mind or orientation of the persons both formulating and implementing the activity. Therefore, there may be serious limitations or some desirable advantages in who is selected to perform an analysis.

There is a common belief, approaching the level of mythology, that all socially aware persons think about safety, health, and medical error in generally similar terms. However, the variations in personal beliefs or opinions are considerably greater than might be expected, even when workplace and personality characteristics are factored out. The following descriptive profiles may suggest what could be expected from those considered for analysis but manifest certain giveaway symptoms or indicators.

Individual Responsibility

There are those who authoritatively voice opinions that all risks should be managed by simply empowering individuals to take more responsibility for their own performance and safety. They assert that individuals should adapt positively to any given situation rather than believing others have some prime obligation for their safety or job. Standards, best practices, and government regulations are believed to be too costly, time consuming, very complex, and far too general to pertain to the discrete unpredictable behavior of any one person. They believe that it is up to the individual, the health care worker, or the patient to be self-reliant, watchful, innovative, resilient, careful, and unemotional in the face of potential harm, error, or disorder. Reasonable levels of risk are to be expected, and some precautions should be undertaken by those in harm's way. They believe that it is unnecessary and unproductive to attempt more than minimal improvements in patient safety. This may not be what the patient or others would expect. With this attitude, little new information can be expected from an analysis except for cover stories to protect the status quo.

Minimalists

There are many specialists or prospective analysts who remain passive or inactive until directed to perform a specific task. They want to rely on a specific procedure, formal technical guidance, detailed directive, trade standard, or government regulation that they view as a target to be accomplished. They need guidelines and are obedient but are not self-energizing in the face of uncertainty. They act as technicians, not professionals. They can discover what is obvious but cannot integrate abstractions, inferences, or diverse factual situations. They are willing to quickly compromise rather than dig deeper to identify additional facts and circumstances. They often obscurate using colorful word games, theories, or specialty fads rather than be factually direct or creative in problem solutions. They tend to be fearful of something new. In essence, they are minimalists in what may result from their analytic endeavors.

Discretionary Agents

There are analysts who seem to enjoy the discovery of facts, personally determining causation, and searching for effective remedies. They generally seek the best-possible countermeasures under the known circumstances. They act reasonably in terms of recommendations, in understanding unique circumstances, and in finding solutions that are appropriate to the situation. They establish goals, can continue their activities without much direction, and can make tentative decisions in areas of uncertainty. They are capable of negotiating compromises if necessary and desirable. They have a personal value system that is consistent with the discharge of professional responsibilities. They understand moral, ethical, and liability considerations. They can select among various analytic techniques to utilize the best for a particular situation. They can provide truthful and effective results based on their analytical skills and past experiences.

Caveats

Choice

There are a wide variety of disciplines with varied methodologies, techniques, and operating philosophies that could be utilized in attempting to resolve medical error problems. It is a matter of choice and careful selection to identify and utilize the type of analysis best suited to correct or prevent the kind of errors manifested.

Reaction

The very process of conducting an error investigation makes others aware, focused, and motivated in terms of their job performance. This generally produces a mild and temporary positive error reduction effect. Whether the overall reaction is positive depends on the skill, sincerity, honesty, singleness of purpose, respect for others, and diplomacy used.

Persistence

Some error countermeasures may be quick fixes or ad hoc interim measures pending further study, equipment modification, or funding for a change. Some countermeasures may be obviously temporary and subject to fading with time, such as stating the need for closer supervision, remedial education, or close attention to the tasks performed. In an error analysis, the search should be for countermeasures that are persistent and remedies that foreclose the possibility of error.

Self-Help

The use of trained in-house teams may provide the best and most economic error reduction efforts. There may be knowledge of the culture, bias, subject matter, and reputations of others in the organization. This may bias the results, prejudice perceptions, or create interpersonal difficulties. Some degree of independence is required, whether in the conduct of the inquiry or for supporting the conclusions.

Economic Evaluations

Estimates of the monetary values associated with residual risks may include monetary "benefits" that could be derived as contrasted with monetary losses or "costs" that have been predicted. Economic evaluations are dependent on various assumptions regarding variables that change with time, place, relevant events, discount rates, offsets, extrapolations, growth rates, and other calculations. These estimates are often manipulated to achieve a desired result. The exercise of diligence and caution can easily prevent major mistakes.

Risk Assessments

The qualitative and quantitative techniques used for risk assessment serve to force a detailed examination, analysis, and categorization of risks. Estimates of risk, with or without adequate data, should include both a reasonable worst-case and central or reasonably expected value. There may be many assumptions and subjective opinions, which could generate an iterative analysis and design improvement process. Good management attempts to convert the unknown and uncertain into whatever can support informed decision making and error avoidance.

Fuzziness

Not all facts are precise, certain, and nonvarying as some analyses and computer models suggest. The human brain can analyze, reason, and infer from imprecise data, partially true facts, vague assertions, and nonlinear functions. Fuzzy logic helps to adapt to such uncertainty, aids in organizing data, helps to manipulate varying logic systems, can deal with partially true variables in continuous systems, learns about rule patterns from limited inputs, and provides values between the ones and zeros of binary logic. The use of fuzzy logic may seem to complicate an analysis; it initially depends on the expert's choice of rules and generally requires an ongoing accumulation of data. To some degree, the fault tree analysts and others do moderate their inputs, and adaptable equipment does adjust to varying conditions. However, considerable caution should be exercised in dealing with fuzzy or defuzzicated outputs, rules, controls, and mathematical functions or equations of cause and effect.

Application

The use of appropriate analytic techniques may provide a second opinion or a first independent assessment of risks. In the past, the analytic emphasis has been on medical device design for consumer safety, medical equipment for safety in hospitals, and manufacturing process safety for drug production. There are new challenges with open innovation policies and the comparative advantages of combinations of sources for research, design, development, materials, production, distribution, and a diverse supply chain. Analysis should be applied to all stages in the evolution of agents useful for countering drug-resistant, vaccine-resistant, foodborne, and nosocomial bacterial diseases. In addition to evaluating safety, risk, and other factors, there should be a direct focus on medical error and its early prevention.

Time Frames

There are increasing competitive and financial pressures that are compressing the time frame for the design, development, and production of devices and drugs. Some overlap of functions may be beneficial if there is not too much rigidity in old and orderly divisions of labor, skills, and functions. Analysis activities could promote the needed flexibility, responsiveness, and cross-discipline accountability now essential. Any excessive time compression that shortens normal activities may produce overt or hidden errors unless there is concomitant independent analytic evaluation activity.

6

Human Factors

Human factors deal in part with the interaction between a patient or person and medical devices, medical equipment, primary care and specialty physicians, hospital staff, and others in the surrounding health care environment. This behavioral interaction is a prime source of medical error. The application of common sense to human factors problems often seems to be a futile effort applied to problems that keep reappearing. What is required are research-driven, evidence-based facts and proven techniques that can be appropriately applied to the prevention of human error. The subject is not new. Human factors have long been regarded as the leading source of anesthesia-related complications (Cooper et al. 1978; Flin et al. 2003). The term *medical human factors* has been used by human factors research specialists working in the field of medicine. Physicians are "feeling overworked, overwhelmed, and underappreciated" (Green et al. 2005), which suggests a behavioral propensity for errors.

Hospital Beds

A federal guidance document on hospital bed entrapment illustrates how human factors research data, guidelines, and preferred practices can be applied to a health care problem. During a 21-year time period (1985 to 2006), there were 691 reports to the Food and Drug Administration (FDA) indicating that 413 people died, 120 were injured, and 158 events were near misses in terms of patient entrapment ("Hospital Bed" 2006). Entrapments are now reported as adverse events to the FDA database. Entrapments occur in hospitals, nursing homes, long-term care facilities, and private homes. In hospital beds, there are seven entrapment areas or zones: within bed rails, under the rail between rail bed supports, between the rail and the mattress, under the end of the rail, between split bed rails, between the end of the rail and the side edge of the head or foot board, and between the mattress end and the head or foot board.

Life-threatening entrapments usually involve the head, neck, and chest. The openings in a bed system should not allow the widest part of a *small head* to be trapped. This guidance document used the fifth percentile female head breadth (120 mm) and, sometimes, the first percentile dimension. The prevention of *neck* entrapment considered neck compressibility, loss of muscle mass with age, and the asymmetrical shape of the neck. The first percentile female neck diameter of 79 mm was reduced to 60 mm to account for tissue compressibility. The angle to prevent wedging (by V-shaped openings) was established as greater than 60°. The *chest* dimensions used the 95th percentile male chest depth of 318 mm. Tests were developed to ensure compliance with the intent of these dimensional measurements, and an instructional video was prepared to demonstrate hospital bed measurements.

To prevent errors of entrapment, there should be additional clinical assessment, monitoring, and attempts to meet the needs of vulnerable patients. Clinical guidance documents, a *Guide to Bed Safety* brochure and a *Guide for Modifying Bed Systems and the Use of Accessories,* were prepared. An attempt was made to involve a wide group of organizations in the formulation of the recommended bed dimensions to ensure widespread agreement of the consensus type. However, New York has adopted more stringent dimensional limits for hospital beds. The FDA has recommended that health care facilities conduct a risk-benefit analysis to ensure that the steps taken to mitigate entrapment risks do not reduce clinical benefits or create unintended risks. Errors may still occur from worn components, damaged rails, softer mattresses, mattress overlays, or replacement of bed system components.

Anthropometric dimensional data are available in publications of the Association for the Advancement of Medical Instrumentation, the National Center for Health Statistics, the U.S. Consumer Product Safety Commission, the International Labour Office, the U.S. Air Force Systems Command, the Society of Automotive Engineers, and architectural publications (such as Watson et al. 1997).

Fatigue

Characteristics

Many physicians became aware of fatigue and the resulting errors and near misses it caused during their hospital residency (www.apsf.org). The long hours and work overload may have produced physical and mental symptoms such as a feeling of tiredness, cognitive carelessness, and perceived need to rest or take a restorative break. The subjective feelings of fatigue refer to a sense of weariness, lack of energy, and reduced motivation. It is accompanied by decreased mental alertness, impaired job performance, increased sleepiness, a higher rate of accidents on and off the job, and falling asleep on the job. There are studies that showed that some anesthetists "do not fully appreciate the debilitating effects of stress and fatigue on their performance" and have "attitudes suggesting invulnerability to the effects of stress and fatigue" (Flin et al. 2003).

Causes

Fatigue is a nonspecific symptom with many causes, including long hours at work, sustained stress, and physical exhaustion. It may be symptomatic of clinical depression, a manifestation of grief, a result of nutritional deficiencies, a side effect of medications, or part of an anxiety reaction. It could be from prolonged physical or mental exertion, extended physical discomfort, continued sleeplessness (insomnia), or physical and mental abuse. It may be part of an endocrine disorder such as hypothyroidism or a result of an infectious disease. Treatment options correspond to the diagnosed cause of the fatigue.

Prevention

In terms of psychosocial fatigue, prevention includes less than 8 hours of work each day, ample rest and work breaks, the use of sleep as a partial cognitive shutdown, and reduced workload. Shift work adjustments may achieve better harmony, rather than disruption, of physiological reactions, social rhythms, and family life. Coffee is often used for a wake-up effect. It has a psychoactive effect that lasts 3 to 7 hours and increases the dopamine and adrenaline (epinephrine) levels. Fatigue may follow as the caffeine wears off. Amphetamines have been used (Berkow et al. 1997; Konz 2006) to reduce fatigue. These include methamphetamine (speed), which interferes with the reuptake of serotonin (a neurotransmitter). Abuse of the drug is far too common. There may be special training to recognize the limitations of human performance when the effects of fatigue begin to appear. Social support has a positive effect under certain circumstances. Job monotony is included in the ILO (International Labor Organization) mental fatigue allowances (Konz 2006). High-stress situations should be moderated or reduced in intensity. Prevention options should be selected after medical and ergonomic assessments.

On-Call Problems

Hospitals that are licensed to provide emergency services are staffed by emergency physicians to provide service 7 days a week, 24 hours per day. The physicians must be available within 30 minutes. The hospitals must also have a roster of specialty service physicians available for on-call care or consultation to comply with the 1986 federal Emergency Medical Treatment and Active Labor Act. About one-fourth of all emergency care cases require the services of a consulting medical or surgical specialist (Green et al. 2005). In a state like California, the on-call consults total more than 2.5 million per year.

An on-call surgical specialist may receive a call at 2:00 a.m., go to the hospital, perform what is required, and then return home for an abbreviated sleep session. Shortly after this, despite lack of sleep and a feeling of fatigue, his office hours may require his presence at 9:00 a.m., to see the patients who are scheduled. A few days of evening on-call services and the specialist's normal schedule may result in diminished performance and a higher risk of medical errors associated with fatigue. The specialist may neglect office patients.

Some on-call specialists are resentful, frustrated, and simply unwilling to take calls because they receive little or no money for it (Green et al. 2005). But, their medical staff privileges may be conditional on such on-call services. Thus, on-call response may be mandated despite the fact that there may be problems collecting payment of down-coded fees from health plans or from earmarked indigent health programs. Reimbursement rates may be so low, perhaps 50% or less, that it is not worth the time and administrative expense to collect them (Green et al. 2005). Some health care plans send patients to an emergency department, then refuse to pay because the services were rendered by noncontracted physicians. There may be cherry-picking by selecting patients who can pay but sending uninsured patients to other health care providers. Such compensation-derived problems can lead to attitudes and behavior that create opportunities for near-miss mistakes or medical errors. Some hospitals may now guarantee payments, provide on-call stipends or per-case payments, or otherwise take measures to ensure the availability of on-call specialists.

Many on-call specialists prefer to have more time for a balanced lifestyle, family interests, or a more flexible professional practice. However, we have interviewed medical specialists who seem to enjoy a near-total time commitment to their profession. One specialist owned one clinic, supervised another, and had daily commitments to three hospitals, explaining that he wanted to accumulate enough money to pursue his real personal interests. His work hours were long, but he was alert, attentive, and responsive and did not manifest any symptoms of fatigue. There are individual differences.

Fatigue and pressure to increase output in a weapon-dismantling project nearly caused the detonation of a W-56 nuclear warhead ("Mishap" 2007). An unsafe amount of pressure was used in the disassembly process. The technicians were sometimes required to work 72 hours a week. The plant's operators were fined $110,000 in November 2006. The destructive power of the weapon was a hundred times that of the atomic bomb dropped on Hiroshima.

Work Shifts

A grand jury investigated the death of a female patient in a New York City hospital and found that unsupervised and fatigued residents were a contributing cause. This led to a revision of the state hospital code, effective in 1989, limiting the work shift hours. Hospital resident hours were limited to 24 continuous hours and a weekly total of 80 hours averaged over 4 weeks. Clinical tours of duty must be separated by at least 8 nonworking hours. Senior attending physicians must provide contemporaneous and immediate supervision of house staff. Hospitals must review each resident's credentials and specify the clinical privileges, such as ability to perform a specific procedure alone, under the general supervision of an attending physician, or only under the direct supervision and observation of the attending physician.

In 2003, work hour limits for first-year residents were recommended by the Accreditation Council for Graduate Medical Education. The 30-hour rule limited interns to a maximum of 30 consecutive work hours, the 80-hour rule prohibited more than 80 hours a week (averaged over 4 weeks), and the 7-day rule provided 1 day free from work each week. In 2006, a nationwide study reported that *83.6% of interns did not comply with these standards* (Landrigan et al. 2006). Some senior physicians disapprove of work hour limits because they believe fatigue does not present a threat to patient safety.

A study found that interns suffered twice the injuries at night than they did during daytime (Ayas et al. 2006). The day following an overnight shift they suffered 61% more needlestick and sharp object injuries The most common causes of work medical errors were reported as lapses in concentration and fatigue.

Other professions may enforce work hour limitations with better results. For example, commercial airline pilots on domestic flights are limited to 8 work hours each 24 hours (U.S. Federal Air Regulation 121.471). International flights must have three pilots, without specifying the job rotation, for flights up to 12 hours (U.S. Federal Air Regulation 121.483). This allows pilot napping to reduce fatigue. The insistence of following the domestic 8-hour rule may result from the actions of labor unions since the airlines could reduce costs by longer work shift hours.

Defiant Actions

In a computer-wired hospital, there always seems to be some medical specialists who are reluctant or defiant in the use of new technology. They prefer paper to computers. They seem indifferent to better infection control. The benefits of a computer system, including drastic reduction of medical error, now seem obvious. But, some people are slow to adapt, change old habits, or learn something new. This is despite the fact that hospitals and most medical specialties are subject to constant change. This defiance of change is important since a fragmented medical informatics system is inconsistent with any future national health system and the present objective of virtually eliminating most forms of medical error. Some defiant behavior may be a manifestation of a psychosocial problem.

There have been many campaigns to reduce or eliminate hospital-acquired infections. The infections may persist because of a similar reluctance to comply with the rules by evasive or defiant actions. Staff may not wash their hands or use antibacterial lotions sufficiently. Some physicians visit their patients in a hospital dressed only in business suits. The patient may be physically contacted by a staff member without gloves or appropriate garments such as lab coats, scrubs, or other protective gear. Hospital staff members may be seen walking on public sidewalks or going to local restaurants while dressed in lab coats with a physician's name tag attached as a badge of honor. The question arises regarding the function of the garments. Are they to protect a member of the hospital staff or the patient? Do they serve in actual usage to gather and transmit bacteria from many sources to many patients? Are these openly defiant actions or errors a manifestation of some psychosocial problem?

Both patients and health care staff may present symptoms of various personality traits and disorders. About 2% to 16% of the general population exhibit symptoms akin to an oppositional defiant disorder (American Psychiatric Association 2000). These symptoms include persistently testing limits, ignoring rules, or exhibiting stubbornness or unwillingness to negotiate or compromise. However, most psychosocial reactions are borderline and nonpathological. These benign reactions may be troublesome to others attempting to reduce medical errors. The relationship between human error and personality traits or conduct clusters (disorders) is described in detail elsewhere (chapters 5 and 9 in Peters and Peters 2006b).

The costs of defiance are huge. Each hospital-acquired infection may cost $40,000 to $50,000. Malpractice is a threat. Consolidation of facilities is occurring with more stringent rules to obey. Despite suggestions to the contrary, based on the fact that health care insurers pay for treatment not prevention, insurers still expect the elimination of hospital-acquired infections as a reasonable goal in the near future. More sophisticated insurance loss control equates with lower insurance costs for compliant institutions. Comparisons between hospitals, long resisted, may soon occur.

Stress

Occupational stress is commonplace among medical specialists, who feel some personal responsibility when facing demands that could result in a patient's good health or permanent disability. The life or death decision-making tasks may act as stressors that produce tension, anxiety, high blood pressure, fatigue, and disruption of the regulatory functions of the human body. Some stress arousal might be helpful, but severe and prolonged stress could produce physiological and psychological changes that increase the likelihood of judgment mistakes, behavioral errors, and poor overall human performance.

When there are unusual stressors, some people remain calm and cool, alert and responsive, with no apparent stress reaction. Others become emotional, troubled, anxious, frustrated, and with deep feelings of social defeat. Job stress coping behavior or styles vary widely, sometimes affected by nonworkplace (external) stresses, including family and personal problems. Stressful life events may result in stress-induced alcoholism or excessive ethanol drinking. Above a certain blood alcohol level, there can be an abundance of human error, as suggested by automotive vehicle codes prohibiting driving under the influence of alcohol. Industrial hygienists remind us that heat stress can kill, and that we should be concerned about stressors such as chemical exposures and musculoskeletal or physical stresses. As stress increases, thought processes suffer and attention span narrows (Sexton et al. 2000).

The countermeasure to stress-induced medical error is primarily to limit and control stressors. Provide adequate safety factors so that errors do not occur before recognition of the unhealthy stress reactions. A focus should be on the accommodation of all reasonable variations in human behavior that might occur under conditions that could foster an adverse stress reaction sufficient to produce medical error. Provide social support to reduce high perceived stress. Widen decision latitude and reduce job demands. Provide appropriate medications. Stress management training may be advisable.

There may be an acute, short-lived stress reaction or disorder, an adjustment reaction or disorder with an anxious mood, or other stress reactions that result in unexpected medical error. If there is a posttraumatic stress reaction, then the stressor stimuli associated with the event may trigger flashbacks or a reexperiencing of the traumatic event. Often, there is difficulty concentrating on job tasks, resulting in errors of omission.

If stress is defined as a resultant of factors that cause bodily or mental tension, then stress may be a cause of fatigue or some other disease entity. Those stress-inducing factors should be identified and controlled, consistent with the physical and mental demands of the job, task, or project. A balance should be achieved since underloading results in boredom, inattention, and unnecessary errors of omission. Overloading, beyond a short period of time, could result in difficult and stressful demands such that human performance suffers and errors are the consequence. This is particularly true when conflicts arise with established work patterns, there is a reliance on short-term memory, and careful monitoring of events is important. In essence, managed stress is important for preventable error reduction.

Because of its importance to medical error, a more precise understanding of stress might be helpful. Stress may be initially created by the reaction and activation of the brain's primitive limbic system that controls emotion. The limbic system includes a memory-processing function that can inhibit or minimize a stress response that is not justified by reviews and comparisons with past experience and present stimuli. If a stress response is justified, then the locus coeruleus in the dorsorostal pons (adjacent to the brain stem), produces norepinephrine, which prompts arousal and sends stimuli through its efferent projections (neurotransmitter system) to specific areas of the entire brain (particularly the forebrain). This stress response activates an opium-like system to reduce pain sensations and a dopamine system to produce a sense of euphoria (the risk-taking "high"). A coeruleus dysregulation contributes to cognitive dysfunction and judgment errors.

The physiological basis for stress involves the interaction of the hypothalamus in the brain, the pituitary gland, and the adrenal gland. The hypothalamus releases the corticotrophin-releasing factor; it triggers the pituitary to release the adenocorticotrophic hormone, and this causes the adrenal cortex to release stress hormones, including cortisol. The cortisol increases the availability of the carbohydrates, fat, and glucose needed to respond to stress. Cortisol is a corticosteroid hormone; if elevated too long, it will adversely affect muscle tissue, decrease inflammatory processes, and suppress the immune system.

The psychological effects of stress are increased arousal, awareness, and alertness. There is increased vigilance, watchfulness, and carefulness. The heart pumps, blood flows, and muscles become energized until the crisis is over. Associated with high levels of stress are panic attacks that may include chest pains, breathing difficulties, and depression. Confronting a highly stressful situation, some people just freeze and become nonresponsive. Continued stress produces high blood pressure, loss of appetite, weight loss, muscle wasting, and gastrointestinal ulcers. There can be stress sensitization, so that subsequent stressors can produce an overreaction of the stress response system.

In terms of prevention, personal recognition of stress reactions and the subsequent development of an action plan may be helpful. A low level of stress may be error reducing. A higher level may invite pharmacological intervention, but drugs such as amphetamines or cocaine may cause sensitization to new stressors. Prolonged stress may require psychotherapy, particularly if it leads to depression with disrupted sleep patterns, high levels of anxiety, dementia, or the other disease entities.

Situation Awareness

The mental process involving perception, knowing or comprehending elements in an environment, understanding of what is going on, and predicting how things will change in terms of relevant information is known as *situation awareness* (Endsley 1988). It involves the dynamic update of a situation using attention, perceptual processes, the working memory, and the executive function of the brain. It is the process of determining what is happening in a system context, identifying and observing relevant variables, and predicting what will happen in order to make adjustments or sound an alarm. Failure to detect and correct in a timely manner may be a situation awareness error.

A decrement in vigilance may be associated with slower and more incorrect responses. In some situations, errors may start to occur as soon as 30 minutes into a continuous vigilance task, and this escalates after 2 hours. The number of errors depends on whether it is a high-intensity vigilance task and whether there has been sleep deprivation or sleep debt. Vigilance errors may account for as many as half the reported cases of avoidable anesthesia mishaps.

An anesthesiologist was servicing two patients simultaneously, each in an adjoining surgical suite. One surgeon requested that the alarm on the patient oxygenation monitor be turned off because there was an interference with the reception of a baseball game he was listening to on ear buds. The inactivated low oxygen concentration alarm failed to sound, and the anesthesiologist failed to detect the oncoming problem in terms of situation awareness. The patient suffered cardiac arrest sufficiently long enough to cause brain damage.

Far too many medical specialists believe that they are good multitaskers, but the human brain functions essentially as a single-channel mechanism. Distractions during complex mental activity can result in selective attention (alert focus), a narrowing of the perceptual field, and comparative blindness to other salient information in the visual field. There may be selective activation of the lateral intraparietal area, modulated by the working memory and interpreted by the prefrontal cortex. In essence, an inattentive or distracted person is prone to errors of omission.

Training of the medical specialist on the limitations of situation awareness may be complicated, as exemplified by the multitasking of automobile drivers, who dial and talk on cell phones while driving in complex mixtures of vehicles traveling around roadway curves, attending to traffic signal lights and pedestrian walkways, and weaving around slower vehicles. Such drivers may firmly believe that they can safely multitask, and the same beliefs are encountered by medical staff members who believe that they can do more than one thing at a time. Perhaps they can do multiple simple things, particularly if they have developed mental expectations from prior experiences, habits, or customary practices. This does not apply to error-free performance in complex situations. Special training may be necessary to prevent situation awareness errors. Such training should emphasize the single-channel limitations of brain function, the necessity of maintaining human vigilance over time-extended complex procedures, and the complexity of situation awareness.

Anesthesia simulators have utilized complex physiologic and pharmacologic models to determine the simulated patient's responses (Gaba et al. 1995). Critical events and faults that require corrective decision making can be inserted by the instructor as part of research or training exercises. The anesthesiologist may be busy with many routine tasks when atypical, subtle, or marginal cues may be introduced into the situation for problem-solving responses. Evolving situations may require a situation assessment, a reevaluation of priorities, and a new situation awareness for the revised importance of variables. It is important to note that the behavior of those evaluated on simulators is different in terms of situation awareness because they know that they are under study and possibly videotaped. Situation awareness training, for individuals or teams, should reflect actual events that can occur in the reality of the hospital operations.

It is important to understand that the subject matter diversity, interpersonal cooperation, and cognitive intensity of some surgical and treatment teams are among the most complex of human endeavors. Situation awareness in such situations is highly dependent on the prefrontal cortex and associated areas in the midbrain and basal ganglia. The prefrontal cortex guides behaviors toward goals in a focused, coherent, constantly updated, and task-relevant manner. The complex social interaction requires unique social perceptions and social cognition processes. The human factors specialist may believe that more situation awareness is necessary to improve the process, reduce error, and harmonize standard procedures.

In terms of future research, some human factors specialists believe that situation awareness is the capacity to direct consciousness to achieve goals by skilled performance (K. Smith and Hancock 1995). It is considered adaptive and externally directed consciousness that is part of the knowledge-generating purposeful behavior achieved by intentional manipulation. Performance is related to stress, mental workload, and the externally defined goals that generate behavior that samples cues in the situation. In essence, situation awareness is the process of creating knowledge and taking informed actions. This suggests that situation awareness requires clearly defined objectives or goals, identification of important variables, and recognition of known and desired adaptive behavior. This may have relevance to the formulation of simulator training programs for medical specialists.

Reduction and Integration

In the analysis of any system, one typical approach is to break down, fragment, or separate the system into its elements, subassemblies, components, or parts. This reductionism permits a separate examination, study, and evaluation of each segment and its interactions with other segments of the system. Subsequently, the segments can be reassembled or integrated to return the system to its original state. At that time, a complete system analysis can be performed while understanding how the integrated parts actually work or are intended to work to enable the system to function as desired.

Human factors specialists utilize this technique in what is called *task analysis*. It may be to study a work system in something similar to a time-and-motion study. It may be a function analysis to determine what occurs at each step of a process, procedure, or system. It may be predictive during the design of a system to determine the probable human tasks that would be performed and to make design changes to improve human performance and eliminate behavior that could cause errors.

A similar approach is used to determine the sequence and prompting of events surrounding omissions (drugs not administered), repetitions (double drug doses), substitutions (good or bad drugs substituted for that ordered), wrong dose (incorrect dosage of a drug), and incorrect administration (incorrect route for a drug). The technique may be used to determine each step in the process of stocking drugs at a pharmacy, storage of the drugs, repackaging the drugs, adding labels, and dispensing the medication. It could include the tasks, steps, or functions involved in obtaining payments and carrying accounts receivable from public or private insurance plans. This technique may be used to identify error-producing tasks in office administration, payment processes, laboratory operations, imaging and interpretation procedures, and communications.

Physicians conduct patient interviews to compile a medical history that may be comprehensive or may be focused on a particular problem. The patient may reveal what, to the patient, seems important, may overlook key items, may be disorganized regarding various events, or may be excessively brief in the answers. Some patients are relatively silent, some confusing, some talkative, and some disruptive. To obtain a full history, the interview process may be enhanced by asking more detailed questions about potentially important subjects. The patient's story is expanded by guiding the patient into domains that need clarification or to derive deeper meanings for the symptoms. In essence, this is breaking down the components of the medical systems covered during a physical examination, interview, chart inspection, and record review. Diagnostic hypotheses are generated and tested by further questions when a differential diagnosis is needed. The pieces and segments are integrated, a plan is negotiated, there is follow-up, and a therapeutic relationship is established. There may be several views of reality: the clinician's diagnosed disease entity, the patient's subjective symptoms and illness, the patient's family experiences, and possible occupational disability evaluations.

This same reduction of a system to its pertinent elements, clarifying or spotlighting any discrepancies, and reconstruction of the overall system are widely used. We have recommended it for human error analyses (Peters and Peters 2006b). It is used by reliability engineers in their failure mode and effect analysis for medical device design. It is used by system safety engineers in fault tree analysis. It is used by industrial engineers in manufacturing process improvement. In each situation for which the technique is utilized, the analyst brings a different perspective based on training, objectives, agenda, specialized knowledge, and personal skills.

A slightly different approach used in the design of medical devices is to identify the functional system in the way the medical device is to be used in the surgical suite. The total system is the typical surgeon, the future specified device, and the range of patients to be treated. The device's functional outcome configuration is the system to be reduced to available components, modifiable components, and new components to be designed. Each component is studied and tested regarding how it helps to accomplish the functions required by the system. It is the integration of the components that forms the functional system. It is the final approved device, the orthopedic surgeon, the company representative, and the patient that form the total system for the treatment of the patient. The device could be a neurostimulator adapted from a cochlear implant, a knee replacement, a portable medical monitoring device, a hip replacement, an artificial cervical disc, an external defibrillator device, or an implantable artificial heart for patients with advanced heart failure who are not eligible for heart transplant and are unlikely to live more than 1 month (under the Humanitarian Use Device provisions of the Food, Drug, and Cosmetic Act). Thus, systems are the prime focus of the device designer, and they require successful integration of the necessary components.

Patient Handling

Members of the hospital staff who manually reposition, lift, transfer, or transport patients are at risk for musculoskeletal injuries. In the process of pushing, pulling, bending, reaching, supporting, lifting, and exerting forces in an awkward posture, the back of a nurse may be flexed and the torso twisted so that biomechanical stressors impose significant forces on the spine. In fact, nursing personnel are working in 1 of the 10 highest risk occupations in terms of physical injuries. The majority of nurses during their lifetime will suffer a back injury.

There is no one human factors or ergonomics solution to patient-handling problems. This is because hospital patients and long-term care residents differ in weight, height, physical agility, cooperativeness, cognitive deficits, and unpredictable behavior. They may or may not assist in the transfer from bed to wheelchair, during toileting, for repositioning to make up the bed, for dressing, or for bathing. There may be fluctuations in assistance required for tasks, such as applying antiembolism stockings (A. Nelson 2006). These are all human factors problems.

One remedy is the use of mechanically operated, electrically controlled, patient lifting and positioning devices. The equipment or device may provide a total lift or just a sit-to-stand assist or be designed for other purposes for which assistance is desirable. It may be fixed or maneuverable but should be available when and where needed. It should not be burdensome to use but human engineered for simplicity and ease of use by all nursing personnel in all areas of a hospital or health care facility. Devices such as slings should be easily attached and not create new risks. The design of such equipment and devices should include error prevention as a key design objective.

The use of lifting equipment should be integrated into prescribed patient-handling work practices. Ceiling-mounted lifts are preferred over bed- or floor-mounted lifts. Nurses' aides should be trained and available to help. A policy of no (manual) lifts should be instituted and enforced, particularly during busy periods. There should be special training on how to recognize and perform high-risk patient-handling tasks. Some long-term nurses have evolved their own no personal injury agendas and could serve as mentors.

Items such as back belts may actually increase risks through engendering false confidence. Local government guidelines may be relevant for nursing home patient handling. There should be administrative oversight in terms of job assignments, including provisions for teams if heavy lifting is scheduled. Organizations should schedule periodic audits to identify unsafe work practices; audits are preferably conducted by a human factors specialist trained in biomechanics and musculoskeletal injury prevention.

Perhaps some 67% of low back pain could be prevented by eliminating work that requires bending and twisting (*Musculoskeletal Disorders* 2001). Since this also pertains to industrial situations, many new devices have been designed to prevent such injuries. Those new devices might have applicability to acute or long-term health care facilities. A great deal of ergonomics research has been performed on upper body biomechanics. Those factors identified in upper extremity disorders may have relevance for medical errors involving fatigue, pain, discomfort, and functional limitations.

Drug-Altered Behavior

A significant proportion of all people in this country manifest drug-altered behavior. This proportion includes both patients and medical staff. There may be combinations of prescription drugs, illegal drugs, and other agents that affect brain functioning. The result might be errors of an attentional dysfunction, errors of a short-term or working memory deficit, errors of omission from a social anxiety disorder, or errors from other unexpected behavior.

Among the most common drugs are psychostimulants, opioids, alcohol, nicotine, and cannabinoids The psychostimulants include amphetamines, known as speed or uppers. The opioid analgesics, which include codeine as well as heroin, affect 800,000 people who are chronically addicted and some 150,000 on methadone. Diazepam (Valium), used to treat anxiety disorders, acts on the limbic system, thalamus, and hypothalamus. The brain function of different people is affected in a different manner by a wide variety of different drugs.

Thus, normal brain functioning is adversely affected by controlled prescription medications with behavioral side effects, under-the-counter medications, and illegal mind-altering drugs. The effect of such a drug agent is imposed on an individual's personality attributes and variables, psychodynamics, brain structure, and preexisting level of various toxic industrial chemicals.

The human factors specialist performs analyses and evaluations of human behavior with assumption of normalcy for 84% to 99% of the general population (Peters and Peters 2006b). The error-prone drug-altered behavior may be included in this normal population estimate. Therefore, the human error specialist considers a broad spectrum of knowledge about human behavior when determining the causes and prevention of medical error. There are many neuropsychological screening tests, clinical reasoning guides, experimental knowledge bases, and work system design features that are considered. The human adaptive process, motivation, and cognitive capacities are also included in such an error evaluation. In essence, simple assumptions of an average person may not be appropriate in the investigation of medical errors.

Macroergonomics

There are human factors experts that focus on work system design. This includes the study of organizational structure, management, and processes from a modern industrial psychology perspective. These individuals investigate organizational structure in terms of formalization, complexity, and integration (Hendrick and Kleiner 2002). The emphasis is on human-centered interface design, humanized tasks, and the organization's sociotechnical characteristics in a top-down and bottom-up perspective. One of the benefits is reduced accidents, injuries, and illnesses assumed to be from human errors. It is expected that the designs of work systems will be participatory, dealing with the workers' feelings and attitudes, with decentralization of decision making.

There is a strong belief that people should participate in the planning and control of their work activities. They should be given "sufficient knowledge and power to influence both processes and outcomes to achieve desirable goals" (Hendrick and Kleiner 2002). This belief is utilized to implement change and for problem solving, and it forms a basis for how an overall work system is designed. This method also uses laboratory experiments, field studies, field experiments, organizational questionnaire surveys, interview surveys, and focus groups. The concern is with both overall work systems and problem areas such as work-related musculoskeletal disorders. The National Research Council/Institute of Medicine report on musculoskeletal disorders identified nursing homes as having a high incidence of disorders due to lifting and moving patients (*Musculoskeletal Disorders* 2001). Macroergonomic specialists could provide an overall system analysis approach to all functions and work activities in a hospital setting.

Team Training

The traditional approach to medical education was individual learning in an authoritative setting where passive patients were told what to do. Basic skills might be learned and practiced directly on patients. Residency training involved graded responsibility. Team training involved a caste-like social hierarchy. Errors were just part of the learning process. However, there have been dramatic changes or improvements in medical knowledge and procedural skills, medical equipment and devices, pharmaceutical efficacy and safety, higher social expectations from the medical profession, informed consent, animal rights, the intrusion of regulatory and legal concepts of social responsibility, complex payment schemes and practices, and rapid growth of medical organizations that stress highly interactive group behavior, financial outcomes, and business goals. There are restrictions on the use of human cadavers, live animals, and patients for training, testing, or research. Residency work hours have been shortened. Individual training opportunities for young physicians have been reduced because they cannot practice new skills on patients, even with a patient's explicit consent (Aggarwal and Darzi 2006). Thus, training needs have changed, particularly for team training.

Basic individual skills may be learned using inanimate models and task-specific checklists or video guides. Basic team training can be learned on simulators that provide a risk-free form of clinical education, particularly for difficult procedures. This might provide a virtual reality that could facilitate the acquisition of transferable skills and the reduction of intraoperative errors. Trainees can refine their skills in a safe environment. The cognitive processes or strategies behind surgical tasks can be studied by eye-tracking technology. This form of training could enhance operating room efficiency while identifying medical errors during teamwork, determining remedies, and reducing the error frequency by the practice of appropriate procedures.

Training is often indicated as a remedy if there has been significant medical error. It is also important as medical enterprises encounter forced adaptation to a flood of new and sometimes-complex technology, as the information age requirements arrive, as business models change with mergers and different payment modes, and as audits and oversight functions provide a more open competitive environment. It is true that medical error is often associated with trial-and-error on-the-job self-education. This contrasts with conventional training, which is directed toward specific goals, includes basic skills and essential knowledge acquisition, and is provided by learned trainers in a somewhat formal setting.

Proper training does work. After an audit of intrapartum-related stillbirths in the United Kingdom for which more than 75% had evidence of suboptimal care, educational intervention was instituted (Draycott et al. 2006). Training courses were introduced on obstetric emergency training, and the results were evaluated. There was some focus on the use and interpretation of cardiotocographs and the recognition of fetal compromises, inappropriate actions, and unnecessary delays. They sought to avoid infants becoming physiologically depressed, requiring prolonged resuscitation, developing hypoxic-ischemic encephalopathy, and developing cerebral palsy and cognitive disability. The result was a sustained improvement in neonatal outcome that was considered clinically important. Other training programs have reported subjective improvement in staff confidence, although the objective is improved clinical outcomes. Team training has been proven in the airline industry (Sexton et al. 2000).

To better achieve the error reduction potential of training, some more economic and relevant training approaches have been devised. For example, in the operating room, surgeons (often perceived as the client) and anesthesiologists (often perceived as service providers) may not work as cooperatively as desired, and antagonistic interdisciplinary relationships might occur. Although the best interests of the client should be paramount, is that client the surgeon, or is it the patient? Elsewhere, there may be intradisciplinary personal conflicts within the nursing staff. In an effort to improve teamwork, some medical schools start by having medical students and nursing students work together in a separate facility as part of their training. They may use computerized mannequins to present various symptoms, and they can act in exercises that teach through simulated scenarios. They learn how to think logically and critically, perform as part of a team, deal with miscues and misfortunes, and practice the kind of decision making they will later experience in an emergency room, during neonatal resuscitation, or elsewhere. Training health professionals together, rather than separately, may help each discipline to understand better the other's role and expert capability. The goal is team endeavors in which people work together cooperatively and freely communicate with others to achieve common goals.

Another approach to the teamwork problem is the use of techniques from other sources. This generally occurs in the social psychology of medicine, for which there are concerns about interdisciplinary and organizational attitudes, developing appropriate safety cultures, improving interpersonal communications, dealing with cross-cultural issues, executing team monitoring procedures, and rewiring professional relationships. This includes the attempts to adopt, modify, and apply the team concepts developed in the aerospace industry. Since 1997, commercial airlines must use the Crew Resource Management technique, which stresses teamwork as a way to reduce human error in complex operations. This approach encourages open reporting with immunity from sanctions if an error is reported voluntarily in the Aviation Safety Reporting System administered by the National Aeronautic and Space Administration (NASA) for the Federal Aviation Administration (FAA). It seems appropriate to utilize the lessons learned from other sources even though aviation flight crews operate in a different system context than an interdisciplinary medical staff facing ever-changing diagnostic and therapeutic goals. However, as medical interventions become more complex and interdisciplinary, the opportunity for error increases, and team training of some sort will become more necessary.

Human Factors Experts

The professional discipline known as human factors emerged about 50 years ago as a result of a need to solve problems relating to human interaction with machines, their surrounds, or environmental context and with overall work systems. At first, it was called applied experimental psychology or engineering psychology when there was a human behavioral emphasis. It was called human engineering when there was a product or system design engineering focus. It has been called ergonomics when the emphasis has been on human work system design. In essence, regardless of name, it is the discipline that performs experimental research on humans to obtain objective or evidence-based data that can be applied by human factors practitioners.

The human factors discipline deals with knowledge of human psychological attributes and variables, physiological reactions and capabilities, and human biomechanical characteristics. It attempts to improve human performance, comfort, safety, health, productivity, psychosocial factors, and quality of life. It is a rather diverse but coherent discipline composed of psychologists, industrial engineers, physicians, and others who specialize in human performance improvement.

Most human factors engineers, including those specializing in macroergonomics, have had advanced training in the conduct of scientific experiments, their quantitative interpretation, and statistical test design. This knowledge may be helpful relative to formulating clinical trials and the controversial "noninferiority" and "superiority" criteria used by the FDA. At present, a noninferiority decision-making criterion, plus a set margin, allows a new drug to be approved even if it is worse than an old competitor. In addition, many older drugs have not been tested against a placebo, so there may be no significant benefit from their use. Research review boards and others may wish to avoid errors by consultation with experimental scientists from the human factors profession.

Human factors experts or consultants are qualified by virtue of their specialized education, board certification, and state licensing. Inquiries can be made to http:/bcpe.org, and further information is available in chapter 7 of our book, *Human Error* (Peters and Peters 2006b).

Cultural Change

A great deal has been said about motivating a change in the safety culture to achieve reductions in medical error. This ranges from the inspirational lectures common in industrial management to educational sessions intended to instill something of a religious zeal for workers to strive for error-free performance. Temporary broad-based improvements generally occur in response to motivational pleas to pay greater attention to job tasks, be careful in all work endeavors, and focus on productive teamwork. Often, there is a gradual relapse, and earnest personal commitments may not ensure perfection in terms of error-free performance. However, cultural change sounds good, is easily scheduled, and could be an economically frugal means of demonstrating that something positive is being done about preventable medical error. Any ostensible improvements probably do not change the status quo in terms of role functions, organizational structure, or available equipment and facilities.

Should motivational seminars or continued education sessions be scheduled in the hope of error reduction, it seems logical that human factors psychologists could help since motivation is part of their research endeavors. This assumes that they perform some preliminary evaluations to provide content of value to a particular group. As specialists in human behavior, their competency should exceed that of other informed lecturers. They should be able to deal with such fads as a Hawthorne-type event and evidence-based caveats as science applications. They may help clarify procedural improvements and the application of preventive error countermeasures.

Sentinel Events

Tubing and Catheter Connections

The Joint Commission on Accreditation of Healthcare Organizations publishes the Sentinel Event Alerts. They often deal with problems that are well known to human factors specialists and for which preventive remedies have been discussed for more than 50 years. For example, tubing and catheter misconnection errors "occur with significant frequency and, in a number of instances, lead to deadly consequences" (*Sentinel Event* 2006). These medical errors occur because functionally dissimilar tubes or catheters can be connected. The misconnection errors include connecting intravenous infusions to epidural lines or connecting infusions intended for intravenous administration to nasogastric tubes or to an indwelling bladder (foley) catheter. Misconnections have occurred with central intravenous catheters, peripheral intravenous catheters, nasogastic feeding tubes, peritoneal dialysis catheters, percutaneous enteric feeding tubes, and automatic blood pressure cuff insufflation tubes. In other words, there is a broad range of medical devices with different functions and access to the body through different routes. The problem of mismatching connectors, in human factors publications, has caused emphasis on the design of physically incompatible connections by the use of varied alignment pins or key ways, different screw collars, and the use of color, size, and shape coding (Van Cott and Kincade 1972). Labeling of high-risk catheters should always be done along with rechecking and tracing to the source of all patient tubes. There are alternative couplings to Luer fittings advertised as preventing unintended connections to intravenous lines. This is a topic on which specific human factors research is needed to accommodate all of the various situations that can occur in a hospital setting.

Electrical Mismatches

Electrical misconnection errors have been reduced, but not eliminated, by features now available on electrical plugs (male prong connectors) and sockets (female receptors). There are medical equipment plugs with a dozen different international patterns of prongs. They vary by pin location, shape, direction, and number. They plug into a physically matching socket to complete an electrical circuit. There are some plug and socket connectors with up to 78 poles or contact positions, usually in a socket plug and shielded pin housing. The plugs may have manual connector locks or automatic locking devices to prevent accidental removal or unintended circuit interruption. They are available in various color codes that comply with the standards of various countries. They may have different forms of screw-on collars, varying footprints, or shaped and marked connector attachment bases. As a general principle, each medical device or critical function should have a unique set of connectors that prevents mismatch errors. Universal couplings should be avoided.

Plugs and Disposable Interconnects

There are medical devices with disposable electronic probes and probe connectors to prevent transferring pathogens from one patient to another (a medical error). There are two connections: the probe to a connecting cable and the cable to a systems generator. The disposable probe connector is used only once. The connecting cable, if used a number of times, would require sterilization (including autoclaving) for each use. The disposable probe could be used with intravascular ultrasound catheters, endoscopes, and minimally invasive instruments. Disposable probes are used in the diagnosis and treatment of cardiac arrhythmias, varicose veins, enlarged prostrates, malignancies, and blocked arteries (Kannally 2007). Since there may be several vendor sources involved in the design of a probe system, special attention should be given to the interconnections to be sure that they prevent misconnections (mismatching errors) with other equipment. Eliminating the probe and connector sterilization process by making them disposable eliminates a major source of error and infection.

Surgical Fires

Another Sentinel Event Alert occurred on June 24, 2006, regarding prevention of surgical fires. The ignition sources include electrosurgical and elecrocautering equipment, fiber-optic light sources and cables, lasers, and sparks from high-speed drills. The fire locations are in the airway, head or face, and elsewhere on or in the body. An oxygen-enriched atmosphere is a frequent contributing factor to the fire. There are many flammable materials in the surgical suite, including alcohol-based solutions, drapes, gowns, towels, dressings, and other supplies used during surgery. Among the risk reduction strategies was to question the use of 100% oxygen rather than air or less than 30% oxygen. Another caution is not to drape a patient until all flammable preparations on the patient have dried. Electrical devices should be placed away from the patient when not in use. Since human factors specialists often deal with complex fire situations, this is another area that could benefit from human factors research, equipment development, and improvements in multiperson procedures.

Unfortunately, some people like to play with fire or gamble and take chances with a potential fire situation. Thus, there may be some predictable deviant human factors actions involved in the emergence of a fire that creates a need to counteract such foreseeable deviant behavior. There should be an attempt to interpose a barrier between a possible ignition source and easily ignited combustible material (fuel). Oxygen-enriched air may elevate the flammability of cotton and rayon fabrics depending on their weave, weight, and thickness. Even flame-retardant material may not be effective as the levels of oxygen concentration increase. Clothing should not be made of nylon, acetate, or other material that will melt, liquefy, and increase the severity of burn injuries. Is there a way to quickly put out the fire? Do the prepping agents and ointments need to be flammable? The basic principle is that any atmosphere enriched above the normal oxygen content presents a fire hazard (Tryon and McKinnon 1969).

Ventilator Problems

Still another Sentinel Event Alert, on February 26, 2002, dealt with ventilator-related deaths and injuries. The majority of the cases involved malfunction, misuse, or inadequacy of an alarm. Other causes were related to a dislodged airway tube, a tubing disconnect, an incorrect tubing connection, or wrong ventilator setting. Other factors included room design that limited patient observation, a lack of response to alarms, alarms set incorrectly, and distractions. Again, human factors specialists have a long history of experimentation, testing, and observation of warnings and alarms. Such research in a hospital setting could be valuable in reducing medical errors.

There were early attempts to prevent medical error in respiratory therapy. For example, a respiratory nurse stated, "To avoid misunderstanding and error, verbal orders are acceptable only in an emergency situation" (Murphy 1976, p. 422). A professor of anesthesia indicated that there are many acronyms and symbols used in respiratory therapy, but there may be "confusion" regarding their meaning among various people (Gish 1976, p. 483). A nurse stated that if any part of respiratory therapy equipment becomes wet, "It is a potential hazard to the patient because of possible contamination from microorganisms" (Nelson 1976, p. 308). Thus, 30 years ago there were discussions of medical errors pertaining to directions, symbols, and infection control.

Mechanical ventilation started in 1530 with a fireplace bellows attached to a tube to the patient's mouth as an aid to lung ventilation. Many different types of positive-pressure breathing machines were subsequently developed. The number of patients requiring respirator care has grown with recognition of the needs of those with chronic airway obstruction, including emphysema and chronic bronchitis. Hospitals now use a variety of medical gases, which may be in the care of the respiratory therapist. Respiratory therapy departments were formed for specialized services relating to respiratory care. The team concept has been fostered. The presence or absence of medical error is dependent on the organization and administration of the respiratory therapy department, its functions and staffing, the training of the therapists, and the location and the type of equipment utilized.

Respiratory therapists are more likely to be present in hospitals than in some nursing homes and in home care situations. The respiratory therapist is qualified regarding the application, adjustment, monitoring, tracking, analysis, and remedying of respirator situations that could lead to serious injury or death. Equipment should be designed for ease, simplicity, and safety of use by those patients who do not have the benefit of the knowledge of a respiratory therapist. That is, the equipment design should be reviewed by system safety engineers and human factors engineers as an assurance measure.

In the industrial workplace, inhalation of harmful airborne particles and fibers may have been reduced by the use of paper face masks or "comfort masks" that workers believed offered some respiratory protection (Peters and

Peters 1997). Gradually, respirators were developed with improved air-filtering capability. The wearability, fit testing, and human factors problems were gradually solved, although residual problems still remain. Where there were toxic vapors present in the workplace atmosphere, a self-contained breathing apparatus was developed. It might have a compressed oxygen tank, a generator, or an air-line respirator from an external clean air service (Ruch and Held 1975). In some respects, there has been a progressive merger of hospital respirator knowledge, industrial respirator protection, and military "gas mask" (chemical, biological, and radiological) respirator knowledge. In dealing so directly with humans in life-threatening situations, human factors problems abound.

Scope of Activities

At one hospital, a human factors engineering project illustrated the scope of activities that can be productive (Gieras and Ebben 2007). This included human factors testing in evaluating medical technology proposed or newly purchased. A user-centered evaluation was conducted of a wireless telemetry monitoring system involving a two-way, voice-activated, hands-free communication badge. The alarm response time was reduced from 9.5 minutes to 39 seconds. This improvement was significant in terms of patient safety, health care teamwork, individual productivity, avoidance of nurse desensitization to frequent alarm pages, and how process improvement can be accomplished on new complex equipment.

Laboratory equipment was evaluated, and graphical user interface improvements were made. The risk assessment process revealed a design flaw that permitted cross contamination of specimens. Orthopedic devices were evaluated for functionality. Incubators were evaluated for safety and ease of use. Tubing misconnections were studied to reduce adverse patient reactions.

It is anticipated that there will be greater involvement of human factors engineers in the rapid advancement of orthopedic and other technology. In corrective orthopedics, signals are sent to the patient to communicate walking with proper balance and posture. If the patient adopts an irregular footstep pattern or an off-balance posture, then a warning signal is given. For prosthetic limbs, sensors and microprocessors can recognize when a patient is losing balance or stumbling and can make some corrections to prevent a fall. However, it is sometimes surprising how the human patient reacts to helpful technology and the patient's refusal to properly care for joint disability, preferring what is believed to be more natural actions or reliance on personal expectations of full recovery. The role of a human factors specialist has become important in fully integrating the electronic, hydraulic, mechanical, pneumatic, and electrical equipment with human capability, reactions, and expectations. In upper prosthetic limbs, including hands, myoelectrical control involves feedback to humans that must be interpreted by the patient as a substitute for the touch sensation. This includes determining whether something held by the hand is slipping and whether the grip needs adjustment. Such feedback to humans may be desirable for many types of implants and disorders.

In many hospitals, staff physicians might see 30 to 32 patients each day at 15-minute intervals. Sometimes, there is considerable pressure to see more patients to increase revenue. There may be a delay of 5 to 10 days for patients to obtain an appointment. To authorize prescription refills, there may be many actions, searching, and walking that consume time. The queue-and-wait, standard-time-and-forced efficiency, and sense of regimentation are not offset by the many advantages in being in a comfortable part of a well organized hospital system. This hurried or rushed process has been blamed for some easily avoidable medical errors. Some physicians have decided that a solo practice or micropractice is better for them. They attempt to convert to a low-overhead, high-technology style that may not even utilize a nurse or receptionist. It is similar to the old, more enjoyable family practice conducted by a lone physician who becomes well known and respected as part of a local community. There are other adaptations in the form of varied types of clinics and even nurses stationed in shopping malls to screen patients. Such situations present many possible human factors questions, particularly in the use of electronic records, computerized calendars, external processing of medical bills, dealing with insurance claims, and e-mail consultations. As the practice of medicine gradually transforms with new discoveries, devices, equipment, and administrative needs, the special talents of the human factors specialist could be of benefit to all participants.

There are an increasing number of surgically implantable medical devices that can monitor key diagnostic functions and send the signals wirelessly via the Internet to a physician's office. These at-home sensors may measure blood pressure within the heart for patients with congestive heart failure (loss of pumping pressure) or hypertension (high blood pressure). They may sense pressure buildup for those with aneurysms (ballooned arteries). They can measure heart rhythms, body temperature, or heart rate. There are questions about how such long-term implantable diagnostic devices could change the behavior of the at-home patient, including food intake, salt consumption, exercise, response to symptoms, or drug dosage patterns. The physician cannot monitor the transmitted information continuously or frequently. A good sampling protocol must be established consistent with a hurried schedule. Human factors specialists have had long-term experience in the design of military system monitors, alerts, information compression, and interpretation guidelines. How would the physician be compensated for the telemedicine activities? What factors should be considered if the treating physician decides to telephone the patient and change drug dosage, diet, or activity in response to the signals received on the monitor, particularly for those individuals with fluctuating symptoms and unobserved at-home behavior.

Human factors research specialists have also had considerable experience in dealing with remote sensing and long-distance control of individuals under varied high-stress situations.

Human factors specialists may be interested in the (human) factors involved in medical error disclosure or nondisclosure both to hospital administrators and to patients. The social, organizational, and professional effects of various types of managerial walk-arounds, surveys, and audits would be of interest. The factors that increase, decrease, or sustain staff attentiveness could be studied. Healthy techniques for countering sleep deprivation and fatigue would be of interest since extended-duration work shifts continue to be a problem despite present guidelines and laws. It is a warning sign when a health care worker falls asleep during lectures, rounds, or clinical activities such as surgery. It may create error when there is insufficient knowledge, information, and warnings about toxic substances such as those covered by the REACH initiative (for the latest REACH information, search at http://ec.europa.eu). Human factors specialists have conducted extensive research on warnings.

In 2006, a director of the FDA spoke about the need to focus on user safety as consumer-use devices become increasingly more complex (Grebow 2006). Recent FDA recalls of some blood glucose meter models renewed consideration of how best to incorporate human factors in consumer devices.

In 2005, an FDA engineer was quoted as saying that more than one-third of all medical device incidents involved "use error," and more than half of device recalls for design problems involved the user interface (Patterson and North 2006). In essence, the devices failed to support the end user. One approach to a human factors analysis was to review the guidance documents, prepare a plan to define the scope of the human factors activity, conduct user research to carefully define the users and the environment of use, perform a task analysis, conduct usability tests, perform a use-error analysis, and conduct a walk-through, talk-through evaluation in the field environment. The human factors plan should vary with the specific improvement needs, but the potential scope of activities is fairly broad and could be incisive.

Caveats

Stress

Before stress reaches the error-producing level, stress needs to be fully appreciated, appropriately recognized, and sufficiently managed. There is often a belief bordering on personal invulnerability and invincibility regarding errors from stress and fatigue. Error-inducing stress and fatigue have many causes, create many different physical and psychological effects, and are modulated by many individual differences in response. Stress control (limiting stressors), stress management (coping behavior), and moderation of job demands (intensity and time duration) may reduce the likelihood of unintended medical error from fatigue and stress.

Fatigue

Long work hours under stress remain a medical error problem. Airline pilots are restricted to 100 work hours a month, truck drivers to 260 hours, shipboard personnel to 360 hours, and railroad crews to 432 hours. The Federal Railroad Administration reported that human error caused 25% of all accidents, and of those, 40% were attributable to fatigue. For example, on June 28, 2004, one freight train collided with another, poisonous chlorine gas was released, and three people were killed and dozens injured. The National Transportation Safety Board concluded that one crew (the engineer and the conductor) had fallen asleep from fatigue. The 80-hour hospital resident work rule, which serves as an example, is routinely violated. Other causes of fatigue are common in health care facilities. Special attention should be given to all causes of fatigue that could result in medical error.

Situation Awareness

Situation awareness involves important vigilance tasks during complex procedures and multitasking and where distractions could occur in a system context. Detection and correction of unwanted excursions in relevant variables may be dependent on marginal or atypical cues. Special awareness training and simulator exercises, particularly as part of team training, may be necessary to prevent medical error. Situation assessments, involving teams and problem solving, are better learned before adverse patient events occur.

Defiant Actions

Involving the disregard or violation of reasonable goal-oriented rules and standard practices, defiant action behavior may constitute a psychosocial cause of medical error. It may adversely affect necessary teamwork, infection control practices, computerization, and on-call emergency services. Professional skills should not always outweigh the probability and consequences of medical error. Administrative actions should be tempered by an understanding of the prevalence, characteristics, and treatment of personality traits and disorders.

Self-Limitations

Whether performed by staff personnel or independent consultants, self-limitations may seriously affect a human factors analysis. Special attention should be given to the analyst's professional biases and personal interests, type of professional credentialing or licensing, research or practitioner orientation, and multidisciplinary capabilities. Many persons claim expertise in human factors, but their ability to perform may be limited or well targeted. Selection for the desired performance is critical in terms of useful results.

Human Engineering

There are many medical equipment errors that have resulted from ignoring fundamental human engineering principles that have been published in textbooks for more than 50 years. This "old knowledge" should be revisited to reduce such easily preventable human errors.

Cognitive Illusions

Some people may seem to be stubborn or irrational and to act contrary to their self-interests. They may not be unintelligent, stubborn, or illogical in failing to adopt new concepts, rules, and procedures or in not making correct causal inferences. They may err because they misread ambiguous cues, misinterpret focused information, apply a valid behavioral pattern to the wrong situation, generally inhibit novelty, use speed in place of tempered caution, or are subject to the tyranny of old mental habits and mentally encoded histories. In essence, there may be cognitive determinants of medical error that can be recognized, countered, and perhaps corrected by psychological means or human factors strategies.

Brand Bias

Well-known products, brands, or logos generally have functional magnetic resonance imaging responses associated with brain areas involved with positive emotional processing, rewards, and self-identification. Lesser known brands activate the working memory and areas associated with negative emotional responses. Direct-to-consumer advertising of pharmaceuticals promotes brand reputation, acceptability, and market demand for the brand. Advertising and brand promotion breed a form of familiarity that could induce a positive bias and a perception gap in decision making. An induced bias could result in unjustified subjective opinions of treatment effectiveness or ineffectiveness. Such bias might be overcome by a greater burden or enhanced credibility of assessed facts and logic. Third-party affirmation or supervisory approval may be utilized to counter such bias, which is not uncommon.

Packaging

Over-the-counter drugs in blister packs could illustrate the intuitive error-free and easy-to-use packaging advocated by human factors specialists. Daily use memory could be reinforced by a 7-day format in a calendar-oriented blister package. Each tablet could remain protected until the peel-push or push-button lidding is actuated or punctured. There is child resistance in the materials selected for the packaging. There is portability, travel protection, and isolation from humidity and water. There is billboard space (packaging real estate) for clear instructions. There may be additional print space on the carton, on inserts, or on information cards and a peel-apart folded instruction and warnings label. There could be photographs of the pill and the pill markings for the identification of the drug. Brand identity and patient safety can be easily enhanced. Package confusion at the point of sale (difficult findability) can be reduced, human readability (and understanding) improved, counterfeiting countered (assurance of nonbogus drugs), sterilization maintained (until the moment of use), and packages bar coded or tagged with radio-frequency identification devices (history and authentication) utilized, whether for prescription or over-the-counter drugs. Some of the specific human factors problems that remain are described in chapter 9 on drug delivery.

Electronics

The workload of the health care specialist may be reduced and performance accuracy improved by some advanced electronics. Equipment can respond to spoken commands or oral inquiries with voice recognition software. There can be text-to-speech guidance for diagnostics, procedures, and instructions. Displays can be combined and integrated and made more intuitive, easier to read, and more accurate when and where necessary. Touch screens can be graphic displays and controls. Specialty laboratories could provide data of clinical utility, with reference ranges and interpretations, to appropriately located displays. However, while such equipment could be or is being developed, it should be subject to documented human factors testing to ensure that causes of potential errors are removed rather than introducing new sources of error.

7

Management Errors

Introduction

The focus may be on medical error caused by individuals, but equally important are management errors. The causal source may be *managerial errors* created by those who supervise and are responsible for a limited number of subordinates. The effect of an error may be limited to a small group. The causal source could be an *organizational error* affecting an entire enterprise, hospital, company, or entity. It could be an *institutional error* affecting an entire professional discipline, a group of affiliated hospitals, a joint venture, a group of partnerships, a merged company with subsidiaries, or a trade group with common viewpoints (see the management chapters in Peters and Peters 2006b). Management errors are more difficult to investigate and remedy since a command-and-control hierarchy has the inherent authority to resist examinations of its decisions and actions, may maintain formal business and social distance, and may not permit perceived challenges in any form or manner. This merely suggests that different tactics, content, and negotiation skills are needed to reduce management error.

Illustrative Error Sources

Managerial

An important lifeline for hospital patients is the call light used to summon help. The call light is also used for minor problems that could be provided by nursing assistants, nonnursing staff, or patient service personnel. There may be a need to reduce call light usage to improve response time and restore the call light to its more important lifeline role (Meade et al. 2006). The nurse manager may instruct all staff members to perform regularly scheduled proactive hourly or possibly 2-hour rounds. Additional rounds may reduce call light usage as well as increase patient satisfaction with having their needs and safety concerns met if specific listed actions are taken during the rounds. Assistance with patient self-care tasks such as ambulating, using the toilet, and eating may require rounds adapted to staffing patterns and patient needs. Rounds may be performed by interdisciplinary teams, by registered nurses, or other nursing staff. In essence, a more systematic approach to bedside rounds is an important patient care management objective.

Organizational

Some 17,000 health care workers "contracted hepatitis B from exposure to contaminated blood and bodily fluids," and 300 died in 1983 (Borwegen 2006). This led to a bloodborne pathogen standard (HR 3839, November 26, 1991), state laws, and the Needlestick Safety Prevention Act of 2000. Hepatitis B vaccination among health care workers now approaches 80%. A question that persists is whether there is an organizational and institutional policy regarding the protection of hospital staff from airborne particle attacks using anthrax, plague, smallpox, or other deadly agents. Using a surgical mask instead of a respirator approved by the National Institute for Occupational Safety and Health (NIOSH) is a violation of the Occupational Safety and Health Administration (OSHA) respirator standard (29 CFR 1910.134, 2006, 419–443). Does the hospital organization have a 4- to 6-week supply of air-filtering respirators, as in other countries (Borwegan 2006)? Does the organization meet or exceed correct federal standards?

There are generally differences in organizational policies regarding disaster plans. Errors can be compounded when there are no rules regarding behavior and actions during a chaotic situation. When there are conflicting obligations, such as disabled parents or children with an acute disease, should nurses and staff report to work as scheduled, or should they voluntarily appear ahead of schedule to provide assistance? How should they participate in postdisaster briefings? Are the policy rules reflected in employment contracts? There should be organizational policies that serve to clarify expectations and responsibilities (*Ethics* 1994; *Code of Ethics* 2001; Chaffee 2006). The policies should indicate what types of disciplinary actions are permitted by the organization. Are there provisions for some employees to be exempt from disaster responses because of special needs? How are employees to be treated if they have a second employment obligation or commitments to disaster relief organizations? Detailed disaster preparedness, policies, and action plans could minimize organizational errors.

Transformational Issues

There some major public health and quality assurance problems that could have transformational effects on health care in the near future. Each problem calls for early proactive management consideration, broadly informed decision making, and long-term plans and actions. The likelihood of management errors during each of the transitions and transformations may be rather high. In part, the problems, as manifested, may be related to concerns about community involvement, social responsibility, appropriate funding sources, the training and retention of sufficient competent staff, and the trend from isolated professionalism to an integrated health care system. The following examples seem to present significant management challenges.

Transitional Clinics

The growth of low-cost walk-in clinics located in large stores and at shopping malls may reduce the uninsured people who are forced to obtain primary care in hospital emergency rooms. There should be a good system of medical record keeping at such clinics, preferably in a generic, limited-access, Web-accessible file rather than an elaborate medical database system if the costs are to be kept low. We have inspected a few storefront clinics, the medical records that were available, and talked to some of the clinic patients. The clinic processes were generally informal and simplistic, sometimes the medical records were cursory, and the clinical acumen of the medical practitioner varied widely. It seemed obvious that there would be more medical errors in the informal setting of the clinics than in a regular hospital that has more formal procedures, supervision, teamwork, and record keeping. It is assumed that these outreach clinics will gradually become more sophisticated, and errors will be reduced.

Nursing Home Changes

Many hospital patients are sent or discharged to nursing homes for rehabilitation, for long-term care, and to increase the hospital's patient turnover. The transfer of patients clears the hospital's beds for other, more acute incoming patients and keeps hospital stays short. The cost of nursing home care is substantially less than hospital care. The nursing home stay is frequently extended, and many patient stays become permanent. The patients often fear a loss of personal independence in a regimented nursing home offering monotonous activities and little free choice. Some patients develop a sense of hopelessness and despair. Both the patients and their relatives may also fear nursing homes because they have heard stories about atrociously poor care and outrageous medical errors. Various state laws have forced some changes in both for-profit and nonprofit care facilities. There are about 10,000 such facilities, and with the shortage of registered nurses, error reduction and upgrades are difficult to achieve.

Nursing homes provide prime caregivers for the elderly. For many years, federal law indicated that financially limited seniors were entitled to nursing home care. Today's busy, mobile, time-consuming society may leave no family members available and capable of giving the needed attention to their elderly relatives, so they reluctantly turn to nearby economical caregiving facilities. Nursing homes may use the patient's pension and Social Security payments to help pay for the health care they receive. They may also receive Medicaid funds. There has been a shortage of part-time, hourly paid, in-home, and assisted living caregivers. There may be restrictions on the number of hours that can be paid to caregivers. There are problems in providing around-the-clock care when needed, providing competent supervision, and achieving error-free medical services.

Nursing homes should be designed to provide care in a home-like, comfortable, and casual setting. Preferably, it is a single-story building with short corridors radiating out from the caregivers' station. It should encourage resident mobility and create a sense of independence, privacy, and convenience. It should not look like a hospital, with long drab corridors and official-looking prominent nurses' stations. Caregivers should have stations close to the resident's room and be able to quickly respond to a wireless digital nurse call system. Ideally, there should be private gardens adjacent to patient rooms. A large multipurpose room and cafeteria or deli should be available to the residents, their guests, and the staff of the nursing home.

There should be adequate fire protection, including fire and smoke detection, containment such as duct dampers to prevent fire and smoke migration, sprinklers in hallways and rooms, ample fire doors, and patient room construction that provides sufficient survival time until trained rescue personnel can remove them from the building. The nursing home staff should be aware of the fire emergency plans and coordinate drills with the local fire department. This may seem idealistic since so many nursing homes do not have sprinkler systems or have systems that fail to operate because of maintenance or water supply problems. Often, a series of mistakes or errors occurs that transforms a minor event into another tragic fire. The avoidance of possible human errors should be included in the emergency fire plans.

For inventory management purposes, a bar code should be placed on each door with an additional bar code for each item in the room. This should be compatible with laptop and personal digital assistant methods of providing detailed inventory reports, financial depreciation reports, and replacement schedules and permit the timely restocking of items. Miscalculation errors and other error problems should be minimized by these computer-based systems.

Such computerized systems could be adapted to verify the dosage, timing, and identity of prescribed drugs along with the patient's identity. For example, in March 2005 it was reported that a 76-year-old nursing home resident was given the medication intended for another resident who had diabetes. The resident receiving the medicine by mistake died of low blood sugar within 1 day. A nurse admitted he had forgotten to check the patient's armband. The resident's physician was not notified promptly. This medical error is not unusual in nursing homes; it occurs at the rate of 1 of every 5 doses in skilled nursing homes (*Archives of Internal Medicine*, 162, 1897–1903, 2002).

In California, each nursing home resident should receive a minimum of 3.2 patient hours of care each day from a combination of registered nurses, nursing assistants, aides, and orderlies. This is one statutory effort to ensure the well-being of nursing home residents and includes the avoidance of medical error.

An incident reporting system using a questionnaire, item list, or menu-driven procedure rather than narrative reports was recommended for nursing homes (L. Wagner et al. 2005). It can improve how the nursing staff assesses residents after a safety incident. The use of bedside computers to record daily medical information about patients was studied by the Centers for Medicare and Medicaid Services in 2003. This was intended to reduce medical errors by electronically tracing a patient's medical condition and daily fluctuations in vital signs (RTI International News Release, April 11, 2003).

Assisted-Living Facilities

Greater attention has been focused on alternatives to nursing care facilities. Assisted-living facilities include in-home care with family members paid to care for aging relatives, elder day care centers, and private boarding homes with small apartments suitable for the disabled. Increasingly, patients and seniors are given the choice of where they want to receive the type of care needed. Medicaid funds may be distributed through a home health agency. The care received in assisted-living facilities may include help dressing, shopping, medical attention, and office visits for therapy.

Family caregiving has become more important as health care costs have escalated rapidly. It has forced families to provide care, including medication regimens and other treatment, to persons with illnesses or functional impairment. Community-based health care is often needed for older persons. There are several problems in family and community health care. First, there may be a long-distance caregiver, one who lives more than 1 hour away. Is there adequate communication and provisions for providing substantial regular personal contact with the person helped and the responsible family member? Otherwise, avoidable error will occur. Second, at the time of discharge from the hospital, the discharge nurse should identify who will be responsible for the patient's care and communicate initial instructions directly with that person. A schedule of subsequent communications should be established, including the best time and day, manner (telephone or e-mail), and alternate persons to contact. Meetings should be arranged that include the caregiver, the provider, the family, and the person receiving the care.

There are approximately 44 million people in the United States who provide unpaid care to disabled adults in their home (Lovely 2006). This is likely to increase because of the increasing cost of nursing homes, increasing health insurance coverage problems, and increasing efforts by the federal government to encourage home care as a way of saving money in programs such as Medicaid. Families may hire temporary help under some programs to help relieve the emotional stress and physical strain from such caregiving. If there are problems relating to patient dementia and the 24-hour care required for some physical incapacitations, then there may be caregiver burnout, financial distress, and family disharmony.

Home Use Devices

There are medical devices for home use that can monitor a patient's blood glucose levels, blood pressure, and cholesterol levels. There are sensors that can, and will in the future, measure much more. The trend is toward increased portability, such as in the form of handheld devices. The objectives are miniaturization, added memory to enable the tracking of test results over time, increasing ruggedness and accuracy, provision of cost-benefit advantages, and creation of all-in-one features. The potential for self-inflicted end-user errors increases with low-voltage and low-current operation capability. For example, patients can walk across carpeted floors during atmospheric low humidity and receive an electrostatic discharge that can cause equipment errors of measurement or other damage. Ease of use from portability is promising, but errors from excessive sensitivity to radio-frequency emissions and other external interference should be minimized. Patients may not be able to understand and avoid the hazards described in an owner's manual, so design resistance should be built into the device.

The manufacture of portable electronic devices is shaped by price competition, performance capability, and customer expectations. There is a trend toward making portable home use medical devices available as nonprescription, over-the-counter consumer products. This increases the probability of human error in the use of such products.

Reliability of Implantable Devices

There are common expectations that there will be high reliability, inherent safety, and targeted functionality for medical device implants that will be in the human body for an extended period of time. Such expectations may be unrealistic given current and projected industry practices. There are often risks beyond those reasonably expected. Some additional management oversight should be given for those who are zealously expanding the frontiers of their professional practice with the use of medical devices. The reasons for caution include the extrapolation or speculative methods used to predict reliability of components in their actual use condition for the duration of their service life. The component may be outsourced and of a relatively new design, without a history or data, including experience in a human patient. It may be a standard component with a history, but the manufacturing techniques or choice of materials may have changed. If outsourced on a cost-competitive basis, there may be some uncertainty regarding how the quality control procedures are actually applied at the supplier's factory. It may be an older component used for different purposes, then subjected to aggressive lean cost-cutting efforts. It may have been designed under the requirements of now-extinct technical standards. Thus, the oral promotion of company representatives should be augmented with more independent objective data. The authorization of management or approval by other peers may be necessary to better control the risks of medical error caused by the poor performance of a new device. The benefits of a particular device, all factors considered, may outweigh the risks, but that should be decided by the management of the surgeon end user as well as the manufacturer.

The reliability of the implantable medical device may be offset by its infection potential from biofilms. Cellular and protein buildup on the surface of a device has occurred on coronary stents, urinary catheters, dialysis tubing, and other products. Drug-coated cardiac stents contain bacterial repellents, but problems remained as of 2007. There has been a lack of adequate evidence concerning the long-term safety and efficacy of percutaneous coronary stents for the treatment of coronary vessel stenosis. In just a few years, this device and procedure was used in over 2 million patients annually. The questions are whether there are advantages over bare metal stents, is it appropriate to use with patients with stable angina, and what are the long-term effects in terms of coronary artery thrombosis and death (Mitka 2006a)?

A minimally invasive disc arthroplasty system to remove and replace the nucleus material with a polymer was intended to restore the gap height between the vertebrae of patients. The surgical tool used was improved to ensure that the desired amount of a two-part polymer was accurately injected. The product improvement replaced pneumatics with a motor driving a ball screw. A magnetorestrictive sensor was reduced in size by 50%. Design problems are resolved after human use for many devices.

While many people would welcome an upgraded, updated, or improved product in terms of its reliability, it should be considered a new product for which data should be gathered. Its use should be predicated on management considerations about the risks, and there should be administrative approval of the new device. The trend is to constantly improve implantable medical devices but to rely on prior history to estimate reliability and safety, thus avoiding substantial proof-of-concept development costs and product development cycles of up to 5 years. This should be an area of management concern.

There may be good reasons why a surgeon would want more specific information about the safety of an implantable device. This may include problems that have arisen with the use of such a device and the need to rule out factors other than individual susceptibility. It may be for better informed consent when questions arise. One approach is to use the adverse event reports submitted to the Food and Drug Administration (FDA) by the original equipment manufacturer (OEM). The OEM should have obtained ample documented information about the component suppliers.

From each supplier, the original medical device manufacturer should have documentation on which management decisions can be based. The supplier should be FDA registered. It should have survived a number of FDA audits on compliance with the quality system regulations (21 CFR Part 820). It should have copies of inspector's logs and FDA-483 listings of potential violations. Warning letters from the FDA and the responses should be reviewed. Documents should be available regarding the quality assurance inspections at each step of the manufacturing process. An important question is, How are violations of specifications, standards, and contracts resolved or corrected? A review of such documentation by the OEM and the end user provides some guidance for management appraisals. There should be more design-oriented information, such as reliability engineering and system safety analyses. This ensures that management considers objective data rather than advertising and promotional advocacy for a particular product.

Disclosures

The disclosure of medical errors has been a controversial subject. It may seem natural for a person to ignore or provide excuses for personal mistakes or errors. In most hospitals, medical errors were investigated and resolved in a highly confidential manner so that coworkers, professional peers, potential patients, and possible litigants would not hear about possibly blameworthy behavior and administrative penalties. The professional's public reputation remained intact and livelihood was not unfairly jeopardized. The assumption was that anyone could make a mistake and should learn lessons from the process of the investigation and possible state report summaries.

There were those who suggested a more open system of disclosures so that there could be more effort in finding remedies, both within a hospital and for hospitals in general. The commercial airline pilot model was suggested as a means of collecting data with better transparency. In the airline model, if the pilot voluntarily reports on an incident, the pilot is excused and cannot be sanctioned or penalized for it. In essence, the pilot is rewarded for disclosure. This has served to encourage timely and more accurate accident reports. It does not reduce investigations; it merely provides a starting point for investigations. The error-reporting system in the aviation industry is known as the Aviation Safety Reporting System and is operated by NASA for the FAA. The Veterans Administration has attempted such a system, and the Institute of Medicine has recommended the creation of a national error-reporting system.

There are significant differences between pilots and surgeons. The pilot is an employee or servant of the airline; therefore, any legal malpractice is the responsibility of the employer, not the employee, under the law. The surgeon generally can be individually sued in tort for personal mistakes, although the hospital may escape liability. Another important difference is that medical patients generally know the surgeon personally, know of the surgeon, or have heard about the surgeon's reputation. Those who fly in a commercial aircraft as passengers do not know or wish to know the pilot or pilot's reputation. Another difference is that one life is in the hands of a surgeon, but a planeload of passengers may be lost in one accident. The pilot may have two or three peer backups, but the surgeon may have none. Thus, there are serious differences in culture, operation, liability, and responsibility between these service providers.

A Patient Safety and Quality Improvement Act (P.L. 109-41), signed July 29, 2005, provided some legal protection to health care organizations if they voluntarily and confidentially report medical errors or submit adverse event reports. Under the Health Insurance Portability and Accountability Act of 1996, data reported to an independent patient safety organization is shielded from legal malpractice lawsuits, accrediting bodies and regulators, and employer actions against health care providers. The data could be used in a criminal proceeding if a judge rules that they contain evidence of a criminal act, are material, and are not reasonably available from another source. A national database is to be developed that is a privacy-protected electronic data exchange and can serve to provide aggregated data reports.

The disclosure problem persists. An adverse event report may have no descriptive content or conclusion. A medical incident may be described in ambiguous terms. Specialists do not want to incriminate themselves even in confidential reports because leaks do occur. Hospitals do not want to prompt third-party investigations that cost time and money and could backfire somehow. Many believe that an intensive investigation is required before all the facts regarding what actually happened become available, causes are determined, and corrections are made, and only then is an error report worthwhile. Some specialists believe that their discipline involves the art of making complex fuzzy decisions, and this should not be held against them. Many believe that others engaged in incident investigation, reconstruction, and cause analysis are familiar with neither their discipline or the patient nor the corporate culture at time of the accident. Medical device manufacturers with international sales may be concerned about how medical error information could be used in countries that comply with social responsibility standards or have applicable corporate manslaughter and corporate homicide bills in effect. This uncertainty may result in a reluctance to disclose anything, even by error, that could connote some form of legal fault. Thus, medical error disclosure, whether self-induced, voluntary, or mandatory, remains a serious public health problem.

Pandemics

The role of hospitals and health care personnel in future pandemic situations should be considered early, assessed regarding possible effective actions, and clearly defined in policy statements. Regional and worldwide exposure to communicable diseases has become more of a threat because of possible bioterrorism, the ease of domestic and foreign travel, the increased density of people in large metropolitan areas, and the vast differences in the effectiveness of public health organizations throughout the world. The questions might be, What action will be required? What effects could transpire if a particular hospital and its staff are faced with a pandemic outbreak that quickly envelops the area surrounding the hospital?

There are lessons to be learned from the 1918 Spanish influenza outbreak and the pathogenic avian influenza zoonotics of 1924, 1983, and 2004. The concern with the H5N1 virus relates to what has occurred to poultry populations and its transmissibility to humans who have direct and prolonged contact with the source of the virus. Generally, humans lack immunity; the fatality rate is about 57%, and the threat is that the virus will mutate and become highly infective and virulent (Ignacio 2006). In 1997, Hong Kong culled 1.5 million poultry to prevent spreading of the H5N1 virus. It helped to stamp out the virus but at a high cost to farmers (McKay 2006).

A new genetically distinct and dominant strain of H5N1 avian influenza appeared in southern China (People's Republic of China) in 2006. It was discovered when the infection rate of ducks, geese, and chickens rose to 2.4% from 0.9% the previous year (Normile 2006). This Fujian strain was found in human cases of flu in China and neighboring countries. Poultry vaccines had little effect on the new strain, and it may have emerged after an antigenic response to widespread poultry vaccinations. As of 2007, there are some promising vaccines.

There have been widely publicized estimates that an influenza pandemic outbreak in the United States could affect 47 million people, cause up to 207,000 deaths, and require up to 734,000 hospitalizations. In addition, there could be 42 million outpatient visits. Perhaps one-third of the population could be affected by the pandemic. This suggests that hospitals should have major worst-case generic preparations for pandemics. How will hospitals protect their staff while providing treatment for those infected? There is currently great uncertainty regarding what should be done, how to avoid panic, and the consequences of possible medical errors.

Among the countermeasures that could prevent, mitigate, or effectively control a pandemic outbreak are the following: Avoid sources of contamination such as H5N1-infected birds, with their virus accounting for 229 human cases and 131 deaths. Stock vaccines, antivirals, and personal protection equipment such as NIOSH-approved N95 respirators. Isolate persons with infectious disease and maintain a social distance of more than 3 feet. Cover the nose and mouth while coughing or sneezing. Utilize proper hand washing and sanitizing. Participate in the community's mass prophylaxis efforts, using preventive treatments, including relevant medications (Sheehan 2006). This includes knowing and accessing the points of distribution or clinics used to store and dispense drugs. What communications will there be with the general public? What are the chains of command with government emergency services agencies? Any hospital plan should be frequently upgraded as evidence-based facts on the effectiveness of countermeasures become available.

There are widespread outbreaks of influenza every year, usually from influenza A or influenza B. The virus is spread by the inhalation of small infected droplets from coughs or sneezes, bodily secretions, or contaminated articles. There are influenza vaccines from inactivated viral strains or particles; these vaccines are changed each year to counter the most prevalent viral strains. There were 85 million flu vaccine doses available in 2005 in the United States. Since treatment is fairly routine and most people can recognize the symptoms of flu, there may be such familiarity that possible pandemics could be downplayed in importance.

Symptoms of influenza C are mild; it has an incubation period of between 1 and 3 days, and the illness persists for about a week. The symptoms are fever, headache, tiredness, body aches, dry cough, sore throat, and stuffy nose. There may be secondary bacterial infections. In England each year, 12,000 to 22,000 lives are lost, and more than 225,000 people will see a physician in 1 week during an influenza epidemic. This flu has constant variations and structural changes. Each year there is a vaccine tailed by the World Health Organization (WHO), and there are other intranasal and attenuated live flu vaccines. Vaccines can reduce influenza-related work absence by 70%, reduce hospital admissions by 60%, and cut mortality by 40% (Dawood 2006). In every step of a vaccine program, there can be repeated minor errors, but the primary errors result from poor and ineffectual occupational health programs. Vaccines should be part of a company's terrorist or natural disaster recovery plan and its key person business continuity plan.

The process of preparing for pandemic outbreaks may be expanded into all emergency security response activities. It is anticipated that emergency first responders on security matters will obtain, in the near future, better detection equipment for biological, chemical, and nuclear agents. The detection and measurements may be at very low levels (parts per billion); they may rely on remote sensing and may involve the process of simplifying and miniaturizing some rather complex laboratory and commercial equipment. Simplification should include error-proof operation and interpretations. There are other potential problems since some biological agents may have an asymptomatic incubation period, result only in weak immune systems, have a sudden massive release, be in the form of new species, or be undergoing a high rate of evolution. Thus, the technology involved with pandemic outbreaks and security responses will rapidly grow under political pressure and become complicated. This should be reflected in hospital planning documents and their updates.

Management could consider some form of a local participatory epidemiology or public health effort. This could be a team for general surveillance, one that could be activated for response and control when there is a perceived threat of a local community pandemic outbreak. The team would gather knowledge of impending outbreaks, probable local hot spots, disease patterns, and how citizen cooperation could be achieved. It could develop a valuable disease control infrastructure for vaccine use after social and cultural factors are considered and with a strong outreach to the community. The local citizens may have valuable information, before and during an outbreak, to contribute to a pandemic participatory response (Normile 2007).

Management Principles

Introduction

There has been an assumption that good management will foster a harmonious culture within an organization, and that group culture should be oriented toward the achievement of common goals, positive social interactions, and voluntarily induced high output or productivity. It can also serve to reduce errors and help in the acceptance of needed changes that could benefit the entire group. There are many diverse management philosophies, and hospitals are considered an exception to the rule in terms of the practices utilized as compared to other endeavors.

Most organizations are still patterned after the old Prussian (Clausewitz) military model in which there is a strict top-down compliance or pecking order. This command-and-control organizational structure is one in which orders are issued to subordinates who are responsible for achieving mandatory performance results in terms of operation, discipline, and administrative matters. There may be strict obedience, habitual stereotypical responses, and by-the-book processes and procedures. In contrast, most hospitals are loosely organized, compliance with policies may be difficult to achieve, and there is a mix of hospital employees and independent contractors (highly trained professionals with allegiance to their discipline). There is de facto decentralization, so centralized leadership, coordination, motivation, and surveillance may be difficult. There are quasi-independent operations that are legally distinct. Central leadership may be rather difficult. There are, however, some applicable management principles and understandable logic for such poorly controlled operations.

Shared Goals

There are some common motivational principles, such as the following: For many years, it was believed that employees would tolerate poor working conditions and authoritative rule if paid to do so. Traditional management practices assumed people prefer to be directed and controlled, that employees disliked work, and that people must be threatened with punishment to work toward organizational goals (McGregor 1960). Starting in the late 1930s, there were studies suggesting that workers formed human social groups with informal leaders to advance the interests of the group (Roethlesberger and Dickson 1939). However, the French philosopher Rousseau in 1762 stated that organizations flourish only when individual and collective interests can be pursued to the fullest extent possible. It took considerable time to fully appreciate the dominance of shared goals as a management motivational principle.

If appropriate, the emphasis is now on formulating meaningful work, job enrichment, self-management, and the opportunity for individual achievement and recognition. There should be a management focus on making personal and organizational goals compatible and meaningful.

Time and motion studies, intended to help support employees by carefully planning each step of their work and removing job obstacles, may have an adverse effect on self-reliant employees. They may perceive such studies as downgrading their tasks, lacking their self-involvement, and discouraging to further job improvement. The adverse effect may promote passivity and errors of omission. The concept of sharing goals includes sharing responsibility for job improvements and occupational enrichment.

Placebo Effects

The Hawthorne studies, initiated by Elton Mayo in 1924, found that when workers were singled out for special attention they would increase their production (what has been called a *placebo effect*). Personal consideration or recognition enhanced performance without other incentives. There was also a social effect; the workers would restrict output to adhere to group norms. Even small changes by one worker may have large side effects across a system. The management principle is to enhance each employee's sense of personal worth, provide a means for recognition, and create ownership or stakeholder interest in reducing medical errors. Fostering placebo affects one employee at a time. Pride in work performance may be an incentive to help prevent errors.

Layering

There is a tendency to add layers of bureaucracy, to establish more deliverable goals, and to create artificial structures to help force progress and achieve many arbitrary benchmarks. As an organization grows in size and complexity, managers usually have less and less knowledge of what is going on, increasingly restricted focus and action boundaries, and less direct operational control. Bloated bureaucracies diminish the role and self-importance of most employees, who may believe they are simply cogs in a machine. An avoidance of layering and hidden niches might be achieved by flat organizational structures or by use of small, semiautonomous organizational units. An isolated and difficult-to-reach executive creates a lack of informed control unless offset by frequent visits or sightings. Professional distancing is similar to layering and fosters medical error. Leadership involves visibility and trustworthy behavior that serves as an unbuffered demonstrable model of desired teamwork intended to achieve known goals. Mergers, acquisitions, and cooperative arrangements between health care facilities and organizations may have to include antilayering measures if they are to retain the desired goal-directed motivation from all staff members.

Management Concerns

Conformance

The management of health care facilities has been characterized by excessive delays in fostering changes in procedures and a slow transition to new technologies. In part, this may be because hospitals provide highly customized services, and the leadership is quite fault tolerant of the actions of the highly specialized professionals for whom they provide services. One recent commentary stated, "It takes, on average, 17 years for the results of a clinical trial to become standard clinical practice" (Porter and Teisberg 2004, p. 65). There have been widespread complaints regarding delays in medical device and therapeutic drug approvals. It is a management concern to balance the delays caused by needed proof of effectiveness and safety as contrasted with the need to provide the latest best treatment options. There should be appropriate predictive or proactive conformance to authoritative rules, guidelines, and recommended practices.

There are many professional associations, organizations, and government agencies that provide guidance information for health care entities. One of the more important sources of guidelines is the Joint Commission on Accreditation of Healthcare Organizations. This group has published National Patient Safety Goals (NPSGs) that highlight problem areas and provide expert-based solutions (goals). Compliance is assessed by the on-site surveys, periodic performance reviews, and the accreditation requirement. Not all hospitals are accredited by the Joint Commission; some argue for longer phase-in periods or a more reasonable time to implement the NPSGs. Some desire an assessment of the degree of implementation of such benchmarks. Note that the very process of waiting for the publication of a goal may be considered tardy, not predictive of required compliance, and a loss of "get ready" time. There may be delays in the form of postponing action until the final date of the required implementation, requesting clarification, protesting, asking for a slower pace, and recommending additional changes and exclusions.

The 2007 NPSGs include improving the accuracy of patient identification by using *two patient identifiers* when providing care, treatment, or services to eliminate wrong-patient errors. There should be a *read-back* of verbal or telephone orders and critical test results to reduce communication errors. There should be a standardized *do not use list* of abbreviations, acronyms, symbols, and dose designations that should not be used for orders and medication-related documentation so that interpretive errors or reading mistakes are avoided. There should be action to improve the *timeliness of critical test results* received by a responsible caregiver if there is too much time between ordering a test and receiving a test report. A *handoff communications procedure* should provide sufficient and accurate patient information. This includes the opportunity to ask and respond to questions, a repeat-back or read-back procedure, and a review of prior patient data *to reduce forgotten or overlooked information*. The number of *drug concentrations* should be limited and the drug list periodically reviewed for look-alike and sound-alike drugs to decrease interchange errors and improve outcomes. Facilities should *label all medications* and containers, such as syringes and cups, to prevent solutions from placement in or transfer to unlabeled containers. The new label should include drug name, strength, amount, and expiration date or time, and this should be verified by two qualified individuals. At shift change, all medications or solutions and their labels should be reviewed by entering and exiting personnel whether on or off the sterile field. Individuals should comply with *hand hygiene* procedures to reduce the transmission of infectious diseases. A *root cause analysis* should be performed on all cases of unanticipated death or permanent loss of function associated with health care-acquired infection. The patient's *current medications* should be compared with those ordered to prevent transition drug errors such as omissions, duplications, and potential interactions. A complete reconciliation list of medications should be provided to the next caregiver or to the patient on discharge. The risk of *surgical fires* should be reduced by education of the staff on how to control heat sources, manage fuels, and eliminate ignition sources. Facilities should encourage *patient involvement*, including families, in their own care. To prevent wrong-site, wrong-procedure, and wrong-person *surgery*, a preoperative verification process to identify mission information, the patient's expectations, and interventions necessary before the presurgical final verification time out should be used. The operative site should be marked in a standardized fashion, unambiguously, and so it will be visible after prepping and draping. Proceed in a *fail-safe mode*.

Note that these examples are meant to be illustrative. They are incomplete. Please refer to the current detailed Joint Commission standards and goals. They are objectives, not specifications of how to accomplish the targets or an indication of what may be reasonably needed to avoid all errors in a particular organization with a defined management philosophy. Implementation of such goals are a management responsibility that should be enforced by monitoring and audits.

Competition

It is assumed that some competition among health care services is desirable because it permits consumer (patient) choice and serves to reduce costs. Hospital and health care management understand that an increased patient flow helps to achieve efficiency and fosters specialized competencies. Since big seems better to management, there has been a trend toward mergers, acquisitions, joint ventures, practice agreements, and collaborations of particular services, costly equipment, and geographic markets. There have been commentaries that competition may seem to have a focus on financial benefits rather than providing better medical services. This is a distinction that should be important for management because of its implications as interpreted by others. The public emphasis probably should be on improving medical service delivery to meet population needs, social dynamics, and probable outcomes in the context of public health initiatives, disease prevention, and health promotion.

Health care competition may reduce rural care, restrict some costly services, ration some services except to those who can fully pay for them, promote what critics may claim as unnecessary surgeries, and reduce emergency room services. A great deal of competition has taken place at the level of health plans, with efficiencies and restrictions serving to pay high salaries, bonuses, and market capitalizations. As recently indicated, "health plans make money by refusing to pay for services and limiting subscriber's and physician's choices" (Porter and Teisberg 2004). The inference is that there may be incentives to use cheaper treatments that may not be as effective as other available alternatives. There may be less money available to encourage underfunded research, provide staff in-house training, and provide education on cutting-edge therapeutic approaches. Thus, management should ensure that there are means to ensure progress and resiliency in providing the best in available services.

Joint ventures may purchase or share the costs of expensive health care equipment or utilize complex technologies that are difficult to operate, service, or maintain. Such collaborative activities may benefit the consumer, but they should not intrude on the antitrust safety zone established by the U.S. Department of Justice and the Federal Trade Commission. The "rule of reason" used is whether the joint venture will reduce competition substantially or will produce desired efficiencies, which should include reduction of medical error as one benefit. The joint venture may enable proper recruitment and training of specialized personnel, which should improve the performance, output, quality, and safety of operation. The formation, implementation, and services provided by such an undertaking should be the result of a collaboration among technical, legal, and management interests.

Electronic Records and Telemedicine

When computerized or electronic medical records are utilized, hospital management should ensure that those involved understand the security, confidentiality, and legal issues involved in the preparation and use of such records. The patient has a right to privacy and confidentiality, and in some states there may be an ownership interest in the records. There should be procedures to restrict access to those who "need to know," to help enforce password policies, and to provide for obtaining confidentiality agreements from staff, vendors, or others having even limited access to the records. There should be a method for the identification and authentication of the users of such a records management system. There may be data encryption, antivirus software, and fail-safe features that prevent information loss. It is important that there be provisions to ensure audit trails of those having access to the system, particularly those who could modify the records in some fashion.

The utilization of telemedicine involves many similar management issues when it is used temporally or for permanent storage as medical records. Telemedicine may be used in medical diagnosis and treatment; for continuing education and health information systems; for laboratory test evaluation, analyses, and reports; or for links to the general public. Computer security and antihacking provisions are important. There have been attempts to provide for a national strategy on the use of telemedicine and a federal government presence to coordinate developments such as payments for telemedicine services, special licensing requirements, and standards for telemedicine devices.

There are state legal requirements that help determine whether computerized medical records are sufficiently trustworthy for admission as court records. The records must have been initiated, maintained, or processed in the normal course of business (they are business records); they were kept according to defined procedures (routine records); and they should not contain uninformed decisions or hearsay (accuracy is ensured). The presumption of business record trustworthiness may be challenged by showing errors, insufficient accuracy, possible altered data, or printouts that do not accurately duplicate or match the original records.

It is a management issue to develop procedures that are in accordance with local law intended to ensure the reliability, accuracy, and trustworthiness of the records for courtroom use. The same accuracy and objectivity is desired by those medical specialists who may have to rely on such records in their clinical practice.

Health Middlemen

Health care costs have risen to about 16.5% of the total national economy (Wessel et al. 2006). There have been numerous attempts to control and reduce such costs. A wide variety of health care middlemen have emerged and added considerably to costs, controls, bureaucracy, paperwork, and payment problems. There have been discussions about reductions in the "unnecessary quality" of medical services and "too much availability" of services to the uninsured or nonpayers. Hospital managers have been seeking financial health, increased cash flow, contracts directly with suppliers to avoid the commissions and costs of middlemen, a stress on less-costly generic drugs over brand names, creation of open-book rebate-indicator transparency to better understand the middleman's or supplier's financial transactions, contracts for fee-only reimbursements, and creation of cost-control initiatives, while at the same time avoiding micromanagement in the auditing of the revenues and cost cycles. Physicians have been dramatically impacted by this transition of the practice of medicine. Many have changed the mode, location, and type of their practice in response. Some believe that the character of the responses has increased the likelihood of error, such as having to practice in areas in which they feel they have forgotten skills or outmoded knowledge.

The managers of health care facilities face future, possibly dramatic, challenges as they and their health middlemen learn, plan, and adjust to anticipated and highly debated legislative changes. These changes include attempts to provide some sort of comprehensive health care reform, overhaul, or revision. There are attempts to "control" health care costs while ensuring access for all. Universal access includes coverage for the uninsured, immigrant noncitizens, the working poor, and geographically distant rural families. It involves reorganization of the health care insurance market, state subsidies, and federal payment arrangements. There has been an increase in private plans for the self-insured, mandated employer group coverage, and a variety of other proposals and plans that might require new taxes or fees. The trend is toward a single-payer universal health coverage that has a per capita cost, in some foreign countries, that is substantially lower than it is in the United States. It is expected that some managers will process such changes smoothly and successfully, while others may commit management errors that adversely affect the viability of their organizations.

A critical aspect of any health care reform is the fact that there are more than 46 million uninsured people in the United States (Hoffman 2007). Many young adults do not believe they need health insurance; most uninsured are unemployed or poor; the average family premium is more than $2,500 per year (a sum that discourages many people); there are 7,800 uninsured for every registered hospital; and the costs for the uninsured are to some degree paid by state and federal funds (Hoffman 2007). Hospital emergency rooms may reluctantly provide some inexpensive treatment for the uninsured. There may be a markup of at least 8% to cover such uncompensated care. Many physicians do provide charity care, but this is an unfair burden on them. The uninsured is less able to afford prescription drugs, continue treatments, and schedule visits with a health care provider. Obviously, there are management problems in dealing with the uninsured or underinsured. However, if there were heath insurance for this group, there would be reduced mortality, morbidity, and medical error.

Caveats

Risk Control

Managers in both drug and device companies are generally pressured to get blockbuster products to market ever faster, to increasingly streamline the production processes, to further improve workflow management, and to increase profits significantly for shareholders. They are expected to implement new technologies smoothly, effectively, and at low cost. They are supposed to deal intelligently with global informatics concepts, resolve complex issues that evolve from product innovation and drug discovery, and control perturbations that become manifest in supply chain management. Each of these objectives may result in management decision errors, process control errors, or the creation of errors normally associated with changes, modifications, and adjustments. Whenever there are changes, managers should sharply increase risk control activities to minimize the risks, including prospective medical error.

Management Action

A lack of action might be a management error that could compound other sources of error. The relative importance of health care errors (98,000 deaths each year in the United States) is evident when compared to deaths attributed to motor vehicle accidents (43,458), breast cancer (42,297), or acquired immunodeficiency syndrome (AIDS) (16,516). If it is that important, then some specific management actions are generally required both to initiate and to maintain significant reductions in medical error.

Remote Control

The functions of top management have often been accomplished by a personal detachment from basic operations and remote control of an organization. This is changing to a more hands-on operation to ensure quicker responses, more timely leadership activities, and more knowledgeable problem solving. A selective sampling of what might be occurring may prevent disciplinary inbreeding, loyalty subsets, wayward departmental goal setting, and a subculture that fosters medical error.

Error Detectives

There is often a need for quasi-independent error detectives to reduce medical error. These may be individuals authorized to cross organizational borders or hierarchies, in a diplomatic fashion, to determine what is actually happening at various locations or with various functions. They provide a direct feedback loop to management without fact or opinion interception, masking, distortion, blocking, being disregarded, or experiencing excessive delay.

Liability Reasoning

There is a legal obligation to take appropriate action when a health care problem would reasonably endanger others. Most managers attempt to take immediate corrective action, some feign innocence and avoid taking action, and a few others make arrangements to ensure that they will have a lack of relevant knowledge of any risk as presumed liability self-protection. It is better to correct the problem than feign or arrange an assumed liability avoidance maneuver that could poise a sword of Damocles over both managers and the organization.

Management Oversight

Everyone seems aware that there are a few germs everywhere, so infection control may be downgraded and could become routine. It may be misunderstood, such as when hospital protective clothing is worn, together with a name identification tag, to local fast food restaurants. However, even a microbe-free environment may contain large sources of pathogenic bacteria. Management has an educational role, such as stressing the 100 *trillion* bacteria in the human intestinal tract, the resultant 1 million cases of inflammatory bowel diseases each year in the United States, and the immune system suppression (susceptibility to infection) from corticosteroid treatments (Marx 2007). The educational role of management should be supplemented by an oversight role to ensure there are no flagrant transgressions due to misunderstanding or indifference to infection control efforts.

Personal Attitudes

Negative attitudes are an important causal factor in medical error. Bad attitudes can easily percolate down the organizational structure. Managers should be aware of what can be inferred from their statements, appearances, memoranda, and decisions. The negative contagion could spread and affect suppliers, third-party workers, and patients. A radiating positive attitude should be displayed by managers. This includes a civil, friendly, welcoming, and appreciative demeanor. A lack of civility should not be tolerated because it induces similar retaliatory behavior with accompanying disruptive and error-prone responses.

Feedback

The hospital staff may be professional and efficient but ill-informed regarding actual patient satisfaction with the kind of services rendered. Patient surveys should be conducted for feedback communications to help determine possible patient problems. Based on the patient satisfaction survey results, there may be additional goals to be identified and achieved. If there is a communications disconnect between providers and patients, then an advisory panel should verify the existence of problems, view the problems as opportunities, and devise corrective actions if necessary. Self-improvement may not occur without real-life communications from the patients.

Conflict

Communications about drug safety reflect the interests, bias, and current state of knowledge of the participants or communicators. This suggests that conflict and uncertainty will prevail. The factual basis of drug communications progresses from promises to clinical trials, expert assessments, field experience, 18-month FDA reassessments, side effect manifestations, withdrawals or recalls, and substitutions with more advanced drugs. Improved drug safety communications, particularly to patients, would reduce drug errors significantly.

Unpaid Claims

An unpleasant aspect of administrative hospital management is the conflict between medical providers and health insurers. Claims payments are routinely delayed, reduced, or denied based on a complex coding system used to file and reimburse claims. There may be 7,000 five-digit codes, each for a medical service or procedure (Fuhrmans 2007). There are reinterpretations of the code, thousands of changing payment rules, and an army of reviewers attempting to control health costs and eliminate fraud. Insurers may take an assertive stand, initiate procedures that confound the medical provider, and outcode those seeking payment for their services. Providers may utilize costly denial management software systems to navigate the insurance claims processing system, facilitate physician billing, and improve practice management. They are often forced into litigation and may demand patients make payments that should be covered by their insurance. One insurer indicated that mistakes are made with special codes; the errors approximate 2% of the claims, and this results from "human error" (Fuhrmans 2007). The unpaid claim problem has evolved into a multibillion dollar industry, increasing health care costs by $20 billion per year. Unfortunately, the administrative burden caused by each insurer is substantial and requires considerable supervisory attention, adaptive flexibility, and some creativity by management.

8

Communications

Interactivity

One of the more complex areas in the operation of a health care organization is interpersonal information transfer. It may be helpful in understanding this chapter to have a basic mental model of the communication process. One model might be that messages (signals) are human-formulated inputs that can be transmitted (sent) over an organization's communication channels (analogous to the organization's nervous system) to other designated humans (recipient interpreters) with an expectation of appropriate (responsive) action. The messages may be conveyed by voice (oral), in print (written), by displays (electronic), by nonverbal gestures (implicit signs), or by silence when messages could be given (implied consent to continued action). The messages may be conveyed or delivered directly or indirectly (by a third person) by speech, lecture, telephone, fax, e-mail, film, print, or other means. They may be proactive or reactive, advisory or directive, routine or emergency, confrontational or supportive, or in more than one language, code, or specialty means of expression. It may be a fuzzy ambiguous message intended to avoid accountability or a pure spin to avoid responsibility.

Acknowledged Problems

Physician-Patient Communications

High rates of medical errors were found in the health systems of Australia, Canada, New Zealand, the United Kingdom, and the United States (Blendon et al. 2003). Patients reported serious health error rates of 9% to 18%. Poor communications were a problem. This included not following doctor's advice because it was too difficult to follow or because the patient did not agree with it. Patients left a doctor's office without getting important questions answered. The doctors did not ask about their ideas and opinions about treatment and care. The doctors did not make clear the specific goals for treatment. The physicians did not discuss the emotional burden of coping with their condition. One in six adult patients stopped taking prescription medications without their doctor's advice because of side effects.

In terms of communications that deal with care coordination, the patients were sent for duplicate tests by different health professionals, they had to repeat their health history to multiple health professionals, their medical records did not reach their doctor's offices in time for their appointment, and they received conflicting information from different health professionals.

Of course, these are patient perceptions that reflect their (customer) dissatisfaction. In addition, they suggest areas for improved communication processes. Serious areas of patient safety concerns may be revealed.

Multiple Communication Errors

A study found that most medical errors in family medicine are set in motion by errors of communication (Woolf and Phillips 2004). Rather than single acts, there were chains of events, with a first underlying error that progressed to a final ultimate error according to most narrative reports. These were anonymous reports of practice errors from secure Internet sources. The categories of harm from errors were physical injury complications, a heightened patient risk for complications, and psychological or emotional injuries (including frustration and anger). Some of the errors were informational miscommunication that might have been prevented by the use of computer systems, or treatment errors that were preceded by diagnostic errors, and some treatment errors that were preceded by other treatment errors. The study found that physicians rarely report on the emotional, financial, and psychological harms that result from medical error.

System Errors

A study of hospitalized diabetes patients found that 33% of the medical errors causing death within 48 hours of the error were associated with insulin therapy (Hellman 2004). Insulin is regarded as a high-risk medication because of flaws in the medical system that invite failures in communication. A lack of coordination between hospital departments or within office staffs results in inadequate communication with patients and in the education of patients. The study concluded that "in flawed systems, errors are made even when individuals do their jobs correctly." The recommendation was to move away from a culture of blame and adopt a culture of safety. This included blame-free reporting of medical errors, electronic patient records, and widespread information-sharing systems with standardized software. A computerized physician order entry system would reduce errors from illegible handwriting, ambiguous abbreviations, inappropriate combinations of medications, wrong medication, and wrong doses of drugs.

Among the errors were infrequent blood glucose monitoring, clinicians' fear of causing hypoglycemia that led to treatment errors, a failure to treat high blood glucose levels aggressively enough, and inadequate education on diabetes for hospital personnel. Patients need health education that is more flexible. Better communication is needed between practitioners and patients. Racial and cultural hurdles to communication must be overcome.

Laboratory Mistakes

Requesting laboratory tests may be easy, but laboratory errors, difficulties in interpreting test results, and the unfamiliar aspects of some rapid advances in laboratory science may result in a misdiagnosis (Pai 2005). Improved communication between the end-user clinicians and the laboratory pathologists is one of the most important means of reducing what has been called laboratory errors. Clinicians should understand the strengths and limitations of tests as influenced by preanalytic variables (such as time of sample collection and drug interactions) that can affect hematological and biochemical values. Intralaboratory errors may occur, generally at a 3.65% rate, but have occurred at up to a 46% rate in biochemistry reports. Postanalytic errors include faulty transcription and faulty understanding of the laboratory report. Error reduction includes automated testing, comments in reports to clarify a point or to increase the clinician's understanding of the results, and reevaluation of unexpected results. The laboratory reports should be clear, concise, and standardized. Appropriate telephone calls to the clinician may improve communications and reduce errors of interpretation. As stated, "You won't go too far by just communicating with your test tube."

Consumer Beliefs

Five years after the Institute of Medicine report of medical errors, 40% of Americans indicated that the quality of health care has gotten worse (Henry Kaiser Family Foundation news release, November 17, 2004). Such consumer perspectives also include the fact that 68% of consumers believe that health professionals are not working together or not communicating as a team. Some 34% indicate they or a family member had experienced a medical error some time in their life, including 21% who claimed the error caused serious health consequences. Consumers believed that the causes of medical error were workload, inadequate staffing, and poor communication among health care providers.

When asked about solutions to medical error problems, consumers indicated that doctors should be given more time with patients, hospitals should develop systems to prevent medical error, health professionals should be better trained, and computerized medical records should be used for ordering drugs and tests rather than the use of paper records. Patients have taken steps to protect themselves, such as comparing their prescriptions with what the pharmacist gave them, bringing a list of all medications and nonprescription drugs to a medical appointment, calling to check on the results of a medical test, asking about surgical details and recovery, and bringing a friend or relative to ask questions or help them understand what the doctor was telling them (communications).

The Culture of Silence

There is a study that built on prior reports suggesting that communication is a top contributor to sentinel events (Joint Commission 2004), and that 60% of medication errors are caused by mistakes in interpersonal communication. The study explored why people have a hard time communicating even when it may contribute to avoidable errors and chronic problems in health care (Maxwell et al. 2005). The research was conducted using focus groups, interviews, workplace observations, and surveys. The surveys found broken rules (shortcuts); mistakes (trouble following directions and poor clinical judgment); lack of support (reluctance to help); incompetence (a perception that they would never put someone under that person's care); poor teamwork (actions that divide the team); disrespect (behavior that is condescending, insulting, rude, or verbally abusive); and micromanagement (abuse of authority, pulling rank, bullying, threatening, or forcing their point of view on others). The study suggested that these problems are rarely addressed and tend to fester within the organization. It suggested that hospitals should create a culture of safety so that workers can candidly approach each other about their concerns. This report was headlined as "Silence Kills."

Supplemental Techniques

There have been many error prevention techniques described in this book, particularly in the analysis and human factors chapters (chapters 5 and 6, respectively). Reference was also made to the many techniques outlined in the book *Human Error: Causes and Control* (Peters and Peters 2006b). The following additional techniques have been successfully used in studies of communication problems.

Cascade Analysis

Instead of focusing on a single event or error, cascade analysis deals with a chain of events, a string of mistakes, or one error giving rise to subsequent errors. Whenever a defined error occurs, a look upstream should be made to identify probable preceding, proximal, or underlying errors. A look downstream should be made to identify probable final or ultimate errors and consequential injury or damage. There is a strong belief that one error can cause another error or group of errors, and that omitting a cascade analysis may render attempted countermeasures fruitless because of missed causes.

Cluster Analysis

The classification of activities, functions, features, or errors into subsets or clusters that have common traits or features may provide a form of insight regarding cause and remedy for probable errors. There may be successive clusters separated by statistical measures or differences. The examination of clusters may lead to agglomeration (conceptual clustering) or divisive hierarchies (partitioning) as represented on a graphic tree (dendrogram). In studies of social and communication networks, cluster analysis may be used to help recognize communities, clusters, or group activities that can be examined for communalities of error sources and prevention. If an error occurs in one aspect of a group, then it could occur in any of the group.

Network Analysis

Social network analysis reconstructs and maps the relationships, structure, and social interactions between individuals in a defined network. The communication patterns between network modes can be used to identify error causes and likelihood. A variation of the technique is *link analysis* to determine crucial relationships in a network. Related techniques include *critical path analysis, Program Evaluation and Review Technique* (PERT), and *flow assignment*. Network analysis can be used to break down a communications network to determine when, where, and how communication errors occur.

Critical Path Analysis

The widely used critical path technique determines the scheduling of project activities and the identification of the interdependent activities that are time critical and those that have some dependencies on time critical activities. The less-critical or float time activities are not included in the critical pathways. This permits a manager to prioritize activities and determine critical chains and permits continuous supervisory monitoring. This work breakdown structure can be used to investigate and monitor communication channels, networks, processes, and projects.

The use of any technique may be highly productive or just a waste of time depending on the skills, talents, motivation, and knowledge of the investigators. The techniques are merely organized goal-directed specific procedures, not something akin to magic or providing anointed intellectual supremacy. However, without a specific technique even a fairly competent investigator may become disorganized, frustrated, and superficial in resultant findings.

Important Variables

All communications, particularly in a large, complex operational system, should be easily understood, simple, direct, definite, complete, practical, and action oriented. The preferred format may be nonlanguage pictorials, graphics, photographs, schematics, and exploded views. People avoid reading lengthy or complex texts but may skim through a document if there is some reasonable incentive for such an action. The communications should be compatible with the user's expectations, practices, customs, preferences, and habits. Excessive reliance on company representatives for training on how to use the medical equipment is not advisable.

Technical Information

Information vital to the setup, operation, maintenance, and repair of medical equipment should be well documented. This is usually accomplished in the form of owner's manuals, maintenance instructions, design drawings, and troubleshooting guides. There may be safety bulletins, warning placards, recall notices, warranties and disclaimers, material safety data sheets, training advisories, and product stewardship information. This technical information should be in a form that is easily accessed, readily understandable, and easily communicated to others and will last for the duration of the equipment's service life. Critical safety parameters should be highlighted. Such information should have been tested by the manufacturers to ensure that user errors are minimized. Technical communications that are difficult to comprehend, contain omissions, have ambiguities, or have interpretive faults will induce human errors at some time in the future.

Psychological Costs

When a procedural directive is effectively communicated to the target individuals, the compliance may vary according to the psychological costs. The recipient of the communication may perceive little personal benefit as compared to the subjective costs of compliance over the intended duration of the message. The targeted individual may make a subjective risk assessment or cost-benefit analysis, including factors or beliefs such as added workload, inherently burdensome and oppressive, meaningless and nitpicking, ineffective or inappropriate, and so vague as to be unenforceable. It may seem contrary to the attitudes, customs, practices, and acquired knowledge of the individual's peer group. If the costs may seem to outweigh compliance, then defiance of authority may be overt or embedded in conspicuous quasi-compliance.

The personal and organizational benefits of a procedural directive should be appropriately communicated to the intended recipients of the message. There should be a method of enforcement made known and adverse consequences for the noncompliant. That is, sufficient positive information should be supplied to tip the balance in favor of voluntary compliance. The greater the benefit, the longer the message retention will be; the greater the accuracy in responding to the message, the better the resultant morale and the greater the motivation to perfect or even exceed the given requirements will be.

It may seem somewhat contrary to the basic principles of assessing psychological costs and benefits, but humans generally restrain self-interest by the psychological effects of moral and social values. The impact of self-interest is limited by social norms, such as fairness. When competing goals, such as fairness versus self-interest, are present, the brain areas involved and activated are the anterior insula and the right dorsolateral prefrontal cortex (Knoch et al. 2006). The self-interest goals and selfish impulses need to be controlled or inhibited to achieve culturally dependent fairness goals, which are the socially accepted, reasonable, and morally appropriate behaviors. In such a deliberative or decision-making process, there is a strong emotional content from the activation of the anterior insula.

Organizations should attempt to develop and enforce a set of social norms, including reciprocally fair behaviors in which unfair, hostile, or undesirable acts are reciprocated with hostility or penalties, and kind or desirable acts are reciprocated with kindness or favorable treatment. Organizational benefits should be enhanced and personal benefits should be suppressed in favor of social goals. This process may not work with those having psychopathological disorders manifested by excessive selfishness and disobedience of social norms. In essence, in the formulation of procedural directives there should be sufficient information communicated to the recipients that organizational goals can become the preferred social values and objectives.

Customized communications are necessary to encourage and enhance the higher levels of human cooperative behavior that is desirable in a health care organization. A stable cooperative social order requires that some individuals carry personal costs, a payoff disadvantage, for the benefit of the entire group or cooperative culture (Gurerk et al. 2006). If there are too many free riders, those who do not incur a cost to create a benefit for the group, then it will lead to the eventual collapse of cooperation (Henrich 2006). Stable cooperative social orders are maintained by sanctioning noncontributors and norm violators. In addition, high-quality individuals, with good reputations, can stabilize and attract others by some form of publicized contribution. There are long-term payoffs in stable cooperation if there is appropriate communication of that cultural payoff.

The patterns of social behavior and the social interest required in complex human social interaction is apparently mediated by the ventromedial frontal lobe, the orbitofrontal cortex, and the anterior cingulate gyrus. Lesions to those areas might result in impaired social cognition, problems in social interaction, differences in the valuation of social information, and possible sociopathy. There may be a downward effect in social responsiveness and social compatibility under psychological threats and fear stimuli. The anterior cingulate gyrus is particularly important when humans perform tasks requiring consideration of other individuals or the engagement in social interactions (Rudebeck et al. 2006). In essence, normal behavior is dependent on the interaction and integration of several brain areas. Further research would be helpful in determining how to reinforce desirable social behavior, develop appropriate and reasonable social outcome expectations, and deal with more permanent impairments of desired responses to social communications.

Intellectual Factors

Communications between medical specialists often rely on the fact that physicians generally have a high level of intelligence (capacity to learn), a significant repository of specialized knowledge (word recognition), ample creativity (adaptability), and practical decision making (choice behavior). There are generally good reading skills (knowledge acquisition), verbal ability (expression of ideas), a willingness to exchange information with others (inclination to share), good verbal manipulation (reasoning), an ability to deal with uncertainty (stable confidence), and some capacity for error avoidance (rational behavior). However defined, medical specialists occupy a higher plateau of intellectual functioning for communication purposes.

The general population of patients is not so well endowed in terms of communication skills, both in terms of formulating messages and receiving them. With respect to intelligence, this group may be divided into normal and impaired. Normal individuals range from high intelligence levels to a cutoff level for the impaired. Based on a relatively new psychometric classification, using a negative one-tailed distribution, the intelligence levels include the following percentages of the general population:

84.13% normal
13.46% mildly impaired (one standard deviation)
2.28% moderately impaired (two standard deviations)
0.13% severely impaired (three standard deviations)

Beyond the capacity to learn, there are the actual achievements, such as school level graduated, reading skills, comprehension related to social context, and learned coping behavior.

In terms of effective communication, the lower level (not the average) of the target group should be able to meaningfully understand the message conveyed. Otherwise, it is no message at all or a misunderstood message. If literacy is a problem, then oral communications may suffice. Mentoring may become advisable for nonliterate or foreign language employees. Even the obvious may have to be explained or training opportunities provided to the nonliterate to enable mimicking, imitating, or copying behavior. Patients may use concealing, masking, cloaking, disguising, or head nodding to cover their communication limitations. There should be some means to ensure that the intended communications are in fact received, understood, and will be responded to appropriately. The various intellectual factors involved in communications should not be ignored. They are responsible for significant medical error.

Human intellectual ability depends on the complexities of human intelligence as mediated by the prefrontal cortex. This includes staying on task while rapidly integrating new knowledge in a blend of learning and past memory in a process aimed at procedural rationality. The associated subcortical areas in the basal ganglia and midbrain also mediate rapid updates necessary for behavioral flexibility, problem solving, and even the executive function. That is, there is an intellectual neural system for rapidly adapting to the demands of the outside world. Prefrontal control, reward processing, and avoidance of perseveration are important intellectual variables. In terms of external or outside communications, how can neural networks be facilitated in their function by properly encoding the basic messages necessary for the adaptive learning that modifies the desired context and goal structure? High-level cognition must be part of understanding communications; the question is how to ensure error-free compliance with the intent of the message.

Emotional Factors

Communications may be filtered by cultural bias or distorted by the effects of mental health problems. A significant proportion of people, from all walks of life, manifest personality *traits* that may be bothersome. A lesser proportion manifest personality *disorders* that interfere with and may distort communications. In one sense, severe personality problems are distractions that could induce medical error, such as compulsions that interfere with decision making (errors of omission). These particular behavioral vectors, as described in detail elsewhere (Peters and Peters 2006b), may be considered important emotional factors that interfere with the receipt, processing, and evaluation of signals, words, phrases, and messages.

Other considerations are cultural factors that can selectively filter both messages and responses. There should be some concern about subgroup cultures because these practices, beliefs, and customs are learned as children, become internalized, are practiced as adults, and are then passed from one generation to another. They are ingrained socially learned traditions that produce predictable interpretations, opinions, and behaviors in complex social settings. They help to ensure subgroup personal interactions that are harmonious and predictable to the subgroup. It is not wise to conspicuously adapt communications to apparent stereotypes or partisanships. It is better to strive for messages framed in a manner empathic to the person and that can result in emotional contagion or general acceptance within the subgroup culture. The message is adapted, not forced on a subgroup. In terms of organizations, appropriately tailored communications can produce goodwill and favorable responses from those who have no trouble understanding pretested friendly messages. Imposition of directives is always somewhat difficult, so someone in the subgroup should be utilized to help smooth out unfortunate language that may offend, disturb, slur, provoke, insult, or cause an outrage.

Nurse-Patient Communications

Communications between busy nurses and sick patients may seem rather simple, easy, and straightforward because the nurses are in frequent personal contact with their patients. However, nearly half of all patients are not satisfied with such communications. They may believe that nurses often ignore their concerns, have poor listening skills, manifest little respect for their opinions, make them feel helpless, avoid sufficient explanations, and make incorrect assumptions. Other nurses communicate in a manner that brings about a comfortable relationship with trust, rapport, and even bonding. Good communications help medication adherence, emotional well-being, and quality-of-life issues and create respect for the health care organization.

When attempts are made to provide information to a patient, a reciprocal talk-and-listen process should occur. Being a good listener means conveying a sense of understanding, an acknowledgment of patient concerns, and some attempts to take comforting action. In general, helpful information should be provided in accordance with the patient's readiness, capacity to understand, and need to know it. There should be some sensitivity about what information is given, how much detail is provided, when and how the information is conveyed, and why the information is given. It should be within the patient's capacity to respond and in accordance with the desires of the patient, family, and relatives.

Most patients prefer collaborative treatment decisions and relationships with their health care providers. This form of communication is important because "when patients understand their symptoms, uncertainty is reduced, mood may be improved, and thoughts about potential recurrence diminish" (Clayton 2006). The key is that understanding reduces the uncertainties that stimulate anxiety and induce further problems.

Cognitive Dissonance

Communications about facts that oppose our current beliefs, opinions, and feelings are uncomfortable. What is the truth, is it the facts or the opinion? If a person comes to know something and to feel something else, then a state of conflict or cognitive dissonance occurs. There may be covert attempts at reconciliation, adjustment, or harmonization. Such attempts may tend to alter interpretations of dissonant facts, color and filter perceptions of events, sometimes change facts, and often create face-saving explanations. The communication lesson is that facts may be somewhat malleable as interpreted in the face of opposition. Miscommunication may be assessed by the nature of the responsive behavior, by some confirmation process, or by testing communications prior to release. The overall objective is to reduce uncertainty in the meaning of communications.

Records Control

About 40% of medical costs are associated with duplicate tests, erroneous or lost information, and related administrative costs (McWilliams 2006). The communications problem (redundant, lost, and incorrect information) may be due to keeping paper records in a sloppy fashion, in a poorly accessible manner, or in individual physician's offices. Most patients believe they do not have full access to their medical records, and that these records belong to the physicians or the medical system that has possession of them. In some cases, patient information is lost in the records handoff between physicians, clinics, and hospitals (a poor communications process).

Some companies (employers) now utilize a portable digital electronic records system for their employees' health records. The medical records are maintained in a central data warehouse or records repository and are continuously updated. There is access by physicians, pharmacists, clinics, hospitals, insurers, and the patient. One basic principle is that consumers have control over their own records. Products intended for medical use can be bar coded for use in the system. Some companies have established miniclinics dispensing health care services and pharmacies dispensing prescribed medications and other health care products.

There are serious patient privacy issues in such vast open-access communications systems. The resolution of the privacy issues may change the operation of such systems. The promise of dramatic cost reductions and reduction of medical error rates is a very powerful incentive.

In other countries where there is a government-administered health system, there are attempts to further integrate the system to reduce costs and errors, to provide pricing information, and to develop public (online) quality systems or comparative measures of performance.

Signal Detection and Perceptual Set

The human brain resolves information ambiguity and selects signals by making comparisons with mental templates or perceptual sets of prior accumulated information. In essence, the brain may anticipate forthcoming sensory information (predictive evidence), be able to match it to a category-specific representation (perceptual code), make decisions based on the outcome beliefs (active decision making), and act on the relevant cues in a guided manner (response). Studies of brain architecture have found that detection of target stimuli, perceptual inferences, and perceptual sets in the medial frontal cortices and orbitofrontal cortex, with some activation for visual face sets in the posterior region (Summerfield et al. 2006). The flow of information and set-related modulation occurs in the inferior occipital gyrus, fusiform gyrus, and amygdala. There may be an excitation of a fairly extensive neural network, but the frontal cortex transiently codes for predicted representations.

This suggests that there is a communication advantage when future message recipients or targets are informed, alerted, preprogrammed, or somehow sensitized to anticipate and react to specific messages or signals. This is particularly true when there may be ambiguity or potential disregard of important representations, messages, signals, keynotes, sensitive information, or activators. Preconditioning tends to ensure more accurate interpersonal communication.

Communicating Consent

Informed Consent

After the conclusion of World War II, the Nuremberg military tribunals were convened (*Trials* 1947). From this activity, there emerged 10 clauses or canons regarding human experimentation that become known as the *Nuremberg Code* (Peters and Peters 2006b, p. 163). This has become the world standard on experimentation using human subjects and has served to define the acceptable moral, ethical, and legal aspects of such research. The basic principle of informed consent is that each person must consent to his or her personal exposure to risk, and that such consent must be informed, voluntary, and revocable. That is, there must be some reasonable and meaningful communication between the experimenter and the subject of the research, plus free choice on the part of the subject who could be injured.

Canon I on free choice and informed consent reads as follows:

> The voluntary consent of the human subject is absolutely essential. This means that the person involved should have the legal capacity to give consent; should be so situated as to be able to exercise free power of choice, without the intervention of any element of force, fraud, deceit, duress, over-reaching, or other ulterior form of constraint or coercion; and should have sufficient knowledge and comprehension of the elements of the subject matter involved as to enable him to make an understanding and enlightened decision. This latter element requires that before the acceptance of an affirmative decision by the experimental subject there should be made known to him the nature, duration, and purpose of the experiment; the method and means by which it is to be conducted; all inconveniences and hazards reasonably to be expected, and the effects upon his health or person which may possibly come from his participation in the experiment. The duty and responsibility for ascertaining the quality of consent rests upon each individual who initiates, directs or engages in the experiment. It is a personal duty and responsibility which may not be delegated to another with impunity. (pp. 181–183)

The basic principles of the Nuremberg Code were reaffirmed, expanded, and made less discretionary when the World Medical Society adopted the First Declaration of Helsinki in 1964 and the Second Declaration of Helsinki in 1975. An important conclusion was stated as follows:

> The design and performance of each experimental procedure involving human subjects should be clearly formulated in an experimental protocol. This protocol should be submitted for consideration, comment, guidance, and where appropriate, approval to a specially appointed ethical review committee, which must be independent of the investigator, the sponsor or any other kind of undue influence. This independent committee should be in conformity with the laws and regulations of the country in which the research experiment is performed. The committee has the right to monitor ongoing trials. The researcher has the obligation to provide monitoring information to the committee, especially any serious adverse events. The researcher should also submit to the committee, for other review, information regarding funding, sponsors, institutional affiliations, and other potential conflicts of interest and incentives for subjects. (World Medical Association 1964, principle 13)

Voluntary compliance was a failure, and formal ethics review boards were believed necessary (Council for International Organization of Medical Services [CIOMS], 1991, 1993; Council of Europe, 1990, 1996; Belmont Report, 1979; Levine, 1986; International Conference on Harmonisation, 1997). Institutional review boards were recommended to help determine the details of informed consent, particularly when research subjects are from vulnerable groups (Herranz and Ruiz-Canela 2000). The basic elements to be included in a consent form and to be articulated in the consent process were clearly defined (DeRenzo et al. 1997).

Institutional review boards were finally mandated in the United States to review all federally funded human research under the provisions of the National Research Act (1974), the Protection of Human Subjects (21 CFR 50), and the institutional review board (21 CFR 56). In Europe, there was the Good Clinical Practice for Trials on Medicinal Products in the European Community (1990), the International Conference on Harmonization's Harmonized Tripartite Guideline on Good Clinical Practice (1996), and the Clinical Trials Directive (approved in December 2000 by the European Parliament). Therefore, there are appropriate guidelines concerning communication with patients about to undergo medical treatment that could be harmful, and these rules may define medical error in conflict situations.

Unconsciousness

In emergency situations, when the patient is unconscious and unable to give informed consent, should research be conducted on the patient? For example, trauma patients in the emergency room have been given a blood substitute despite the fact that some have suffered heart attacks and were unable to consent. A Food and Drug Administration (FDA) rule permits emergency research in life-threatening conditions in which all available treatments are unsatisfactory. The FDA also grants informed consent exceptions for emergency research (Couzin 2006). The urge to do research and to administer what might be beneficial to a patient and the nuances in interpreting regulations set the stage for what might be considered medical error. The primary and immediate means of preventing error is the manner in which the responsibilities of the institutional review boards are implemented.

Comatose Patients

When a patient is in a comatose or vegetative state, what are the covert or residual mental functions? Is there some communication possible with a patient that is comatose (unarousable), vegetative (no spontaneous intentional behavior), or unaware (awake but no purposeful behavior)? Numerous studies have been conducted using behavioral observations and neurological tests but have been indecisive and subject to gross subjective opinion. A recent functional magnetic resonance imaging (fMRI) study found the existence of some mental life when there are a few cerebral lesions and islands of brain activity, as contrasted with massive structural brain lesions (Naccache 2006). For a vegetative patient, there were speech-specific brain activities found bilaterally in the middle and superior temporal gyri and further auditory language processing in the left inferior frontal region of the brain (Owens 2006). This suggests some speech comprehension function but not that the patient is actually aware of the semantic processing. Mental imagery tasks revealed brain activities in motor areas, the parahippocampal gyrus, the posterior parietal cortex, and the lateral premotor cortex. The authors concluded that a vegetative patient was able to understand spoken commands and respond in terms of brain activity. However, there was no speech or movement response. This suggests that some patients may be able to use their residual cognitive capabilities to communicate their thoughts by modulating their own neural activity. This is an interesting research area, but the debate continues whether any meaningful communication is possible with those who are unaware, unconscious, comatose, or vegetative. It is assumed that further brain imaging studies may gradually clarify whether there could be informed consent of a type that precludes medical error.

The authors of the fMRI study (Owens 2006) are developing a battery of fMRI tests to measure cognitive functions in brain-damaged patients who are unable to communicate (Miller 2006a). With combined fMRI and diffusion tensor imaging (DTI), used to map the tracts of axons carrying information from one brain region to another, a road map can be constructed to help limit neurosurgical damage. In addition, fMRI communication feedback has enabled some people to control their own brain activity in the right anterior cingulate cortex, which manifests pain perceptions. Neurofeedback represents communication between a brain imager and a patient, who utilizes the communication to alter pain perceptions and behavior.

Misreading Symptoms

It has been common to misread symptoms and make a wrong diagnosis. There may be a failure to consider everything that could and should be known. If the correct tests are not ordered, then the correct diagnosis may be missed or delayed. There may be subsequent deficiencies when taking medical histories, performing physical examinations, and developing treatment plans. Misdiagnosis can be a serious communications problem if there is a reliance on inadequate information, premature decisions, and lack of follow-up. Misdiagnosis is not a minor problem; it occurs in 10% to 30% of all diagnoses and is found in the outcomes of 5% to 10% of all autopsies (Landro 2006d).

To provide more relevant information for the decision-making process, there are *decision support systems* that provide a list of possible diagnoses and prompt the ordering of appropriate tests. These electronic online diagnostic aids operate after the patient's known symptoms are entered into the system. The system may then respond with 5, 10, or even 30 possible diagnoses from a possible 10,000 known diagnoses. The prompts may include a series of questions to be asked of the patient to help clarify certain issues.

There may be a problem in communication if the patient sees several health care providers who do not contact each other or make entries into a master electronic medical record file that is available to all. There may be a failure to communicate test results to the patient, who could then communicate the results to others. Some diagnoses are rather simple, but other symptoms and rare conditions can be misdiagnosed. There should be some procedure or system to prevent slipups and overlooked diagnoses, to improve diagnostic accuracy, and to ensure that misdiagnosis can be quickly corrected before serious harm occurs to a patient.

Caveats

Garbling

Poor communications are a major cause of medical error. Inappropriate, insufficient, and garbled messages are transmitted, filtered, and interpreted by recipients with a wide variety of personal attributes, translational abilities, and motives to comply.

Effectiveness

Interpersonal communications are not as simplistic in formulation as is commonly believed. There are complexities that should be considered if the messages and signals are intended to evoke desired reactions without adverse consequences.

Cognition

The brain's deliberative process in decision making should be understood and considered, particularly in terms of compliance or rejection of communications involving procedural directives. Also, see the drug delivery warning section in chapter 9 of this volume.

Informed Consent

A major communications problem occurs with unconscious and comatose patients because of their residual cognitive abilities. There may be two-way communication of some sort, but is it sufficiently meaningful in terms of informed consent, with or without exceptions and approvals?

Misreading

Misreading of information is a communications problem that could result in misdiagnosis. There are effective decision aids and support systems that can help provide better diagnostic accuracy and more acceptable effective treatment.

Beliefs

There are consumer beliefs that the medical error problem is getting worse, that health care professionals are not working as a team, and that a significant percentage of the population have had some personal experience with a harmful medical error. Such subjective personal beliefs may indicate that there is a communications problem in dealing with both those using medical services and those in the adjacent community. Unfair beliefs can be controlling at critical times, so proactive communications may be highly desirable. The problem should be approached directly, not rationalized.

Silence

Most medication mistakes are a result of interpersonal communication problems. Such problems or errors are initiated by behaviors rarely addressed within health care organizations. The virtual silence about real causation begets unending errors. A more fundamental or insightful approach to error prevention may be productive.

Caution

Cascade, cluster, network, and critical path analyses are supplemental error causation identification techniques. They may be rarely used but can be helpful in difficult, complex situations. This includes identification of causation when there are errors in communication systems or networks. The techniques may be applied, in part or in whole, if other efforts have been insufficient.

Nurse Communications

Physicians often utilize nurses to communicate with their patients. For example, nurses may be told to convey test results they do not fully understand, and they may do so from memory or abbreviated handwritten notes. They may not be able to answer the patient's questions, or they commit errors of interpretation because of shallow knowledge of the specific subject matter or the specific patient's condition. Similarly, if the nurse takes certain vital sign measurements, such as temperature, blood pressure, pulse, and weight, the data are often committed to short-term and error-prone memory until entering it on records or a computer file some time later. To avoid memory problems and to facilitate exchanging information with the hospital record-keeping system and with other professionals, nurses could utilize a lightweight, portable, tablet computer. Other alternatives include small handheld computers or computers on wheels. Data can be recorded on tablet computers directly from the medical instruments at the patient's bedside and sent by wireless technology to the hospital record-keeping system. The tablet computer can be compatible with bar code reading, radio-frequency identification devices, and camera imaging systems. Writing data on scraps of paper or on the nurse's hand before returning to a nurse's station to enter the data later is a correctable communications problem.

Fraud, Waste, and Abuse

All health care organizations that receive or make at least $5 million a year in Medicare payments must educate all their employees on how to detect fraud, waste, and abuse. The employer must also establish policies to ensure that their contractors investigate and report abuse. Health care employees will have some whistle-blower protection. The Deficit Reduction Act, effective January 1, 2007, has been characterized as creating a corporate ethics program that could have a significant effect on the reduction of intentional medical error, assuming this new communications channel is open and responsive.

Time Lags

Time is of the essence in communications where delay can cause errors and harm. Time lags can be reduced by more efficient ER patient management triage procedures with separate routes or facilities for those diagnosed. Time delays could be reduced with multipurpose high speed (e.g. 10 second full scan), multi-slice (e.g., 64 detectors), high resolution CT-scanners with images placed directly on the hospital computer system for immediate availability. Time may be reduced by a quick pneumatic tube system between isolated departments (such as laboratories and clinical facilities). Time is saved with direct, immediate, and constant line-of-sight observation of patients. Even quicker availability of hand antiseptics (such as ethyl alcohol) could save time while encouraging infection control. Walls and woodwork without surface properties that assist in infection control may require more time by staff and maintenance. Open visiting hours for intensive care patients may increase family involvement and therefore provide quicker reaction times for patient care. Placement of all patient medical records on the Internet could provide quick reviews by all those entitled to use that information now and in the near future. Management walk-arounds should be in the playbook to assure communication channels that provide for quick problem recognition, immediate correction, and appropriate overall perception. There are many time lags, short and long, that can have a cumulative and compounding effect. All communications should be both quick and accurate to enable fast, reliable, and appropriate diagnosis and treatment without excessive patient flow cues and unreasonable waiting times. Rapid response times (efficiency) should be a cultural mindset that is not burdensome and a sign of professional excellence. Since most physicians have committed blameworthy errors, their confidence can be restored by improvements that are designed to reduce errors.

Modern hospital communication systems may involve self-healing building automation systems that have embedded intelligence and multiple redundant pathways to assure the quickest path around obstructions and disruptions. The concept of reduced time also applies to the dilution and removal of inhalant contaminated air by filtration, displacement, and downward laminar flow in hospital operating rooms, emergency facilities, outpatient clinics in commercial buildings, patient rooms, and doctor's offices. This is more important because of drug resistant, but filterable microorganisms and displaceable contaminated air. The monitored parameters of such breathable air systems should be part of the overall hospital communication system with timely alarms for impending decontamination malfunctions.

9

Drug Delivery

There is a wide disparity between the perceptions of those who supply, directly or indirectly, the drugs, medications, pharmaceuticals, medical devices, and over-the-counter formulations and the perceptions of those who consume, use, or require those products. In addition, the patient or consumer usually has little voice or choice in the prescriptions given, the form of treatment, the container used for drugs, the clarity of instructions, and how specific personal needs can be best satisfied. This chapter presents some of the concerns and problems of the patient as perceived by the consumer.

Containers

Prescription medications may be delivered to the patient, from a pharmacist, in a small container, bottle, pouch, or package carrying enough pills for 30, 60, or 90 days. The label may have little patient-useful information other than the directions "take 1 tablet every day as directed." An older, forgetful patient who usually takes the medication in the morning may stop mid-morning and wonder whether he took the medication that day. A package of loose pills does not indicate whether the patient took the daily dose. If the patient does nothing except worry, then there may be an underdose. If the patient takes another pill to be sure, there may be an overdose. Uncertainty is not desirable for important medications. Some package design help may be necessary for at-home patients with short-term memory problems.

Blister packs, pills in pockets sandwiched between layers of transparent plastic, might help to remove the uncertainty for some patients because there are location cues, formatting, and billboard space. Contraceptives may be day dated, some pills with active agents and some pills as placebos, in a blister pack container. Wallet-style blister packages may provide daily, weekly, or monthly dosage layouts to improve patient compliance. A multiple-panel wallet blister pack may have ample room for inserts to explain the drug's action and to provide information that could encourage compliance.

Packaging that encourages daily use may simply use a 28-day blister pack, with four rows of 7 pills and with the days of the week printed beside the pills. Some pill packs are available in 4-week and 12-week configurations. An information booklet can be used as a folding cover for the blister pack. Some blister packs may need a desiccating polymer insert that can absorb or release small molecules to maintain relative humidity and control the release of vapors and gases. There are also desiccant washers, inserts, bags, and thin films.

Those push-through blister packages with foil lidding may be child resistant. They should be senior friendly, requiring only slight muscular power. They should be anticounterfeiting and tamper evident. The plastic barrier cover should act as a shield against tearing, biting, bursting, and damage from humid conditions while retaining printability (Allen 2006). Wallet-style blister packs can be designed to hold a great deal of information, including directions for use in several languages. Instead of push-out or push-through pills in blister packs, which may be difficult for some patients, there can be a peel-off panel lid or window.

In a hospital-type setting, there may be smart medicine-dispensing carts that arrive at the patient's bed on a schedule; the smart drug labels are automatically checked and compared with the computerized prescription order for the patient, and a printout is given to the patient so that a further check can be made to ensure that the correct medication and the correct dosage were received.

Unit dose packaging may not be appropriate for those with multiple diseases who take many different medications. Some sort out each day's prescriptions into plastic tray compartments to be sure they take everything each day in a simple fashion. They may devise a dosing schedule, sort out the pills by time of day, and place them in separate bowls. Insurance plans may require 90-day bulk purchases for economy. Some use a double-check procedure (such as husband and wife) in setting up their doses. The fact is that too many pills look small, white, and round. When more than one family member uses some round white pills, there could be problems. A better means of visually discriminating between pills is desirable.

Some mixed liquid biologics that have been distributed in the past in glass containers have shifted to flexible packaging or plastic pouches. The emphasis is on ready-to-use premixed solutions that prevent mixing mistakes or errors, particularly for high-risk preparations. The preference is for single-dose packages to reduce possible contamination from partial use and storage. Plastics do not break from minor accidents as glass does. They may have composite barriers to prevent transmission of oxygen and moisture vapor. They can be produced using aseptic processes. Plastic pouches may be used for metered-dose inhalable drugs, liquids, or powders. As the drug is used, the pouch collapses and prevents contaminated air from entering. Pressurization for aerosol delivery is not needed because the thin plastic-laminated walls can be squeezed to evacuate the contents. The plastic pouches should be abrasion resistant, puncture resistant in terms of blunt and sharp objects, tough, and strong. Multilayer films (laminates) may use nylon for strength, then polyethylene, then a sealant. Special plastics may have to be used for resistance to sterilization by e-beam or gamma radiation. Pliable materials for vacuum-formed packaging for orthopedic implants should retain an effective barrier to gas and moisture. If needed, there should be retention of desirable characteristics during cryogenic storage at liquid nitrogen temperatures or for temperature-sensitive drug storage. This is an area in which materials are constantly improved as manufacturers learn of the many unexpected errors that can occur in health care facilities so that the packaging can be made more error proof.

Labels

The contents of labels placed on containers, bottles, pouches, or packages are an immediate and direct form of communication to the patient or end user. It is surprising that so little useful information is delivered for use by a patient compared to information that is directed to others in the supply chain. This may result from the legal preemptions or evidence protections that result from Food and Drug Administration (FDA) label reviews, guidance, and drug approvals. There may be unfounded reliance on supplemental communications from the treating physician and package inserts. The FDA has had an objective of ensuring an acceptable base level of readability and content without excessive costs to manufacturers. Insufficient convincing information on drug labels does lead to patient confusion, ignorance of essential information, and unneeded medical error.

Labels should contain product drug identification, including the drug name in appropriate languages, bar codes that include mass serialization, and radio-frequency identification devices (RFIDs) incorporated into the label design. They can also serve adverse event reporting systems and provide for quicker and more economical recall procedures. The RFID tags are also placed on bottles, cases, and pallets of drugs to provide track-and-trace accountability, reduce counterfeiting, and provide a means of preventing supply chain errors.

An RFID tag placed under the label of blisters, bottles, syringes, and vials may use high-frequency or ultrahigh-frequency radio waves. The items are tagged by the manufacturer; read by electronic readers at wholesalers (distributors) to pick, pack, and ship; and then read again by retailers (dispensers), even if the items are in mixed totes and if cold temperature monitoring is desirable for temperature-sensitive drugs. There is a need for industry standards to ensure interoperability in all of the supply chain. The use of RFIDs permits the accurate collection of data without the need for human intervention and error.

Authentication and anticounterfeiting techniques include the use of invisible bar codes, encoded fiber paper or watermarks, holography, and color-shifting inks. All such techniques can be incorporated into computer-based medical error prevention systems.

To prevent counterfeiting of drugs, some states are urging serialized item-level drug pedigree information. This would be retrieved by RFID real-time electronics and in printed documentation that is human readable and bar-coding readable. By use of a Web-based handheld device, patients could scan bar codes to prevent manual (human) errors in patient data recordings.

The identification and location of those receiving contaminated medications would not be as difficult and error prone.

Note that additional detailed information on label content, warnings, and supplements are contained in the following sections of this chapter.

Warnings

Basic Objectives

Warnings provide a reasonable means and opportunity to provide and communicate risk information to all those predicted or foreseen to be harmed or injured. The warning should appropriately identify the specific hazards and the magnitude of associated risks. The warnings should be in such a form, language, and content that it could provide the targeted individuals with a reasonable and fair opportunity to avoid the harm (Peters and Peters 1999). Warnings provide for informed choice behavior.

This concept is derivative of the informed consent provisions of the Nuremberg war trials (*Trials* 1947), which indicated that a person must consent to a personal exposure to significant risk, and that such consent must be informed, voluntary, and revocable. The target of the warning message must have legal capacity to give consent and have the ability to exercise free power of choice without coercion, constraint, or deception.

Residual Risk

Warnings are generally used only as a last resort after reasonable attempts to remove all significant hazards or side effects that could cause harm. Removal of hazards includes the use of appropriate safeguards to minimize risks. In other words, a product, function, or drug should be inherently safe if reasonably possible. If the risks cannot be eliminated or reduced to socially acceptable or tolerable levels, then the remaining risks should be openly communicated to those who might be harmed.

Compliance

Compliance with warnings should not be countered by user or consumer perceptions that it involves burdensome, inappropriate, meaningless, or ineffective actions. High psychological costs may render the warning ineffective or useless because the costs provide a reason or rationalization for noncompliance.

The individual tends to make a personal risk assessment. If the frequency and severity of injury are believed to be low, uncommon, rare, or remote for a particular user or consumer, then the personal risk may be perceived as low or even nonexistent. If compliance runs counter to custom, practice, acquired beliefs, or prior associated meaning, then compliance may be considered unnecessary. If the risks are believed low, then only a marginal benefit might seem to justify forgetting about compliance with a warning. Once the decision or choice is made, subsequent similar warnings may be overlooked or ignored.

Graphics and Symbols

Symbols and graphics may conserve label space, may tend to overcome user literacy and foreign language problems, may rapidly communicate a message without reading, and may serve as a brief supplement and reminder for other word message warnings. Preferably, they should be internationally harmonized, tested for signal recognition, and have achieved the 85% correct comprehension level commonly used for such messages (*Australian Standard* 1981, ANSI Z535.3–1998).

The color background for such symbols should be in accordance with those commonly used for risk severity levels with which patients may be familiar. The signal word *Danger* is accompanied by a *red* background, just like a stop message in a traffic signal. The signal word *Warning* is accompanied by an *orange* background, just like industrial signs. The signal word *Caution* should be accompanied by a *yellow* background, just like emergency evacuation signs (ANSI A535.1, ISO 7010:2003, ISO 3864-1:2002. A *green* color background is used for *advisory* information, just like a traffic signal light that suggests a safe or "go" situation. Geometric shapes are also used to show severity of risk on a consistent and similar basis worldwide.

Sign Offs

If a warning is needed to alert and inform a target recipient of the potential of substantial harm, then it acknowledges a known risk that will be assumed by others, including the target recipient of the message. In essence, it requests those who are exposed to assume the risk voluntarily or to elect to undertake avoidance measures. In essence, the risk requires informed consent and is somewhat similar to performing ad hoc research on humans. Research may be observational and waiting for possible data in the form of adverse event reports.

In the 1950s, penicillin was not given for the treatment of syphilis in the Tuskegee case, and the patients were merely passively observed. The New Orleans case also involved nondisclosure of risks, observation to collect data, and exposure to different levels of airborne asbestos (Peters and Peters 2006b). The 1962 thalidomide calamity also involved human experimentation without warnings. In 1964, the First Declaration of Helsinki was adopted, followed by the 1975 Second Declaration of Helsinki, the 1979 Belmont Report, and other statements of ethics regarding human experimentation. Institutional review boards were mandated in the United States under the National Research Act of 1974, the Protection of Human Subjects regulation (21 CFR 50), and the institutional review board regulation (21 CFR 56). There were other international efforts to translate ethical principles into enforceable law regarding human experiments (Peters and Peters 2006b).

In research protocols, proof of informed consent usually includes a written document that is explained to each research subject, then signed and dated by the research subjects. In terms of drug warnings, if there is a substantial risk of injury, then it would seem desirable and proper to have written sign offs that indicate the voluntary choice to accept a fully explained and understood risk of harm. This would seem appropriate regardless of the possible benefits to an individual or to society as a whole and whether it is just passive observation or data collection. This is particularly relevant to off-label unapproved use of poorly understood drugs and their mechanisms of action.

Adequacy

The ultimate evaluation of a warning may be made by a juror in a product liability lawsuit. In 2007, a state jury awarded $1.5 million when they found a hormone replacement drug was responsible for a woman's breast cancer (Reuters 2007). The legal theory was a failure to warn, that is, a failure to use clear, proper, and adequate warnings. The company's defense argued that they did clearly warn the users of a slightly increased risk of breast cancer. The plaintiff's attorney argued that the risk was known for decades, but they did not conduct studies to quantify the risk. The lesson to be learned is that all warnings are not equal; they may be evaluated or judged by nonexperts, quantitative data on the magnitude of the risk is important, and postmarket studies should be conducted to further assess the risks and possibly reduce them. The company faces an additional 5,000 lawsuits involving its hormone replacement drugs.

Reinforcement

In an effort to reduce the side effects of a popular acne drug (isotretinon), the FDA initiated a program intended to prevent pregnant women from taking the drug. The side effects included birth defects such as misshapen heads, missing ears, and heart defects (Dooren 2006a). The first warnings were emphasized by having women sign statements that they "understood the dangers." The second wave of supplemental warnings, devised by the four drug manufacturers, was intended to reinforce the earlier warnings. This included having patients register online with the FDA, visit their dermatologist every month for a new prescription, fill the prescription within 7 days, answer a monthly online questionnaire, see a certified laboratory each month for blood pregnancy tests, and be on two forms of birth control. The result was a denial of the drug for some women and general patient frustration with the program. Men must participate in some parts of the program to monitor for side effects such as liver damage, suicidal thoughts, brain pressure, vision difficulties, muscle damage, and blood problems (Dooren 2006a). This restricted access or limited distribution plan permits postmarketing surveillance of the drug to determine the specific risks more precisely, to elect a barrier to misuse for a self-administered drug taken over a 4- or 5-month course of treatment, and to dramatize the risks of the drug for up to a million users a year.

This is illustrative of how warnings can be reinforced. It suggests that a simple warning may be relatively ineffective under some circumstances, such as when there is a strong motivation to use a product regardless of stated risks or when there are understated risks. When warnings must be extensive, a Web site could be used for consumer education if it would pop out when needed. Restricted access, limitations on distribution, registration, close monitoring, questionnaires, and other barriers to use may be needed to reinforce warnings.

There are drugs that might cost $90,000 or more per year, as in the proposed use of a colorectal cancer drug for breast cancer. The unmet need, high price, and potential sales volume of a drug are strong generators of advocacy and induce restricted assumptions of possible side effects or risks. Advocacy may serve as a counterincentive that undermines the perceived need for effective warnings. Some drug advocates may feel that drug sales should not be dampened by earnings that create adverse consumer or patient reactions. Thus, this may be the reason why a product may start with tiny, obscure, and routine warnings of possible risks, graduate to stronger warnings and instructions, and later evolve to various methods of reinforcing the existing warnings.

Mental Processes

For warnings to be reliable and effective at the time and place of need, there are several steps that must occur. First, the design or formulation of the warning must be such that a clear and meaningful message can be conveyed. Second, the warning must be sent or conveyed to the target and knowingly received. Third, the recipient's brain must engage in a learning process involving neural synaptic plasticity. Fourth, the brain must store the memory of the signal, hazard, and response. Fifth, the memory must be triggered, before fade-out, for a timely and effective response to the risk. To properly design or utilize a warning, there should be a basic understanding of some of the more important brain functions that are relevant to warnings.

Potentiation

This brain function involves a progressive increase in the strength of the synapses of interconnected neural networks. This constitutes the basic neural mechanism for learning and memory. What may be surprising is that long-term potentiation is first induced in the hippocampal area (Bliss et al.

2006). In particular, area CA1 of the dorsal hippocampus plays a special role in formation of memories used to avoid or anticipate danger (Whitlock et al. 2006). The strength of the synaptic connections in the hippocampus directly relates to the strength and stability of memory storage and retrieval. Less potentiation occurs during fear conditioning situations, which inhibits plasticity in the hippocampus. Potentiation is an iterative process of learning for distributive association memories. Some of the memories are eventually taken over by neocortical areas during further learning processes, information consolidation, recall, and reconsolidation.

Long-term potentiation, a hallmark of neural plasticity, is typically found at synapses between neurons. However, fast neuron-glia synaptic transmission has been found between CA1 hippocampal neurons and NG2 macroglial synapses (Ge et al. 2006). The rapid neuron-NG2 cell signaling function may allow rapid feedback of neuronal functions.

Subthreshold Reactions

Consciously "invisible" stimuli can influence brain activity, often even more than suprathreshold stimuli (Tsushima et al. 2006). This apparently occurs because (nonperceived) signals may not be subject to an effective inhibitory control. It is an unknowing reaction to the invisible. It is understood that attentional filtering or suppression occurs with weak or irrelevant signals. This suppression apparently does not occur if the stimulus is below, but near, the perceptual threshold. This is consistent with theories that indicate conscious mental life can be governed by memories that are no longer accessible to conscious insight (Stoerig 2006). In contrast, people often fail to notice salient events when multiple stimuli serve to mask or distract. So, they may fail to notice the obvious and are not conscious of it, and it does not consciously guide their behavior.

Complex sequences of thoughts and actions are mediated by the basal ganglia, a subcortical area of the brain. Our ability to communicate such complex information quickly, without major intervention of prefrontal cortex executive functions, is important in facilitating warnings, cautions, and safety advisory communications. The integration of ideas is primarily a cortical function that provides the ability to apply warnings in different or diverse nonrehearsed situations.

The importance of the prefrontal cortex in probabilistic learning and suppression of behavior (response inhibition), plus noradrenaline and serotonin effects on those human cognitive functions, should not be overlooked or underestimated by the warning designer.

Language Control

A vital aspect of how people respond to warnings may be the social context in which danger signals appear. It may involve predicting the mental state and actions of others (empathy) while retaining emotional control (dampened feelings). This social and emotional cognition may involve mirror activities, which are imitations of the behavior of others or simulation activity. This occurs as a left-side brain function involving language circuitry. If there is no imitation, then the medial prefrontal cortex is involved. If abstract rules of behavior are involved, then it is a dorsal medial prefrontal cortex function. The prefrontal cortex serves to control emotions in social situations. In particular, the right ventrolateral prefrontal cortex, by suppressing emotional processes, permits higher cognitive abilities to guide thought and behavior (Miller 2006b).

The head of the left caudate plays a critical role in language control (written words) and semantic content processing (word meanings). The left caudate receives corticostriatal projections from frontal, temporal, and parietal associative regions of the language-dominant hemisphere and connects to the thalamus for selection of motor sequences (Crinion 2006).

Intrinsic Brain Activity

Human functional neuroimaging has revealed that the at-rest brain is engaged in considerable highly organized activity (Raichle 2006). The brain accounts for 20% of the body's energy consumption, about 60% to 80% of the brain energy supports the neuron communication process, but only about 1% of the energy is expended in dealing with the momentary external demands of the environment. In other words, there may be considerable unconstrained spontaneous cognition (external stimulus-independent thoughts) or an active reconsolidation of correlated neural inputs that enhances interpretations of information and the capability of responding to predicted environmental demands. This intrinsic brain function, a brain that is never at rest, permits memory-based out-of-our-own-head perceptions and rapid responses to predicted warning situations.

The at-rest brain generally tends to wander from one thought to another (Mason et al. 2007). This mind wandering or daydreaming is used as a baseline from which people quickly depart when attention is required elsewhere, and it depends on a default brain network with reduced central executive demand. This occurs when the brain is basically unoccupied with specific mental tasks and provides a housekeeping function and an optimum level of low arousal. When fully aroused, the brain spontaneously produces the thoughts, feelings, images, and voices deemed appropriate to the perceived situation. The hippocampus is intimately involved with the memory that enables a person to construct, imagine, integrate, or envision future commonplace scenarios (Miller 2007). Imagination shares the same brain regions, including the hippocampus, in encoding new memories and the cortex for long-term storage. All of the encoding, storage, processing, imagination, and scenario retrieval is involved in determining the effectiveness of a warning for a given situation when the brain is aroused.

Another intrinsic function is loss of memory, particularly with respect to warnings and instructions. Loss of memory may vary from a few damaged synapses to loss of a third of the brain mass (Cookson and Hardy 2006). Cognitive decline due to disuse, normal aging processes, or Alzheimer's disease should be considered in evaluating the target audience for warnings and instructions. There may be synaptic strengthening, leading to memory formation or actions by the amyloid precursor protein, which gradually leads to the deposit of amyloid aggregates or neuronal death. The persistence of memory is a vital factor in the effectiveness of warnings. Drug warnings ordinarily should be designed for a wide range of patients, from those with good memory to those with evidence of some cognitive decline.

Derivative Criteria

Long-term potentiation requires more than a single episodic exposure to a warning signal, hazard description, risk consequence information, and the initiation of prospective avoidance reactions or behavior. Appropriate potentiation may lead to responses to signals of which the person is not consciously aware or that occur during brain rest. Warning response may be subject to filtering, suppression, masking, inhibition, forgetfulness, or fade-out, which may give rise to error. The design of warnings should attempt to overcome or reduce the effects of warning suppression by the mental processes of patients and other people.

A brief exposure to a warning may or may not create a short-term memory sufficient for a desired response. Long-term potentiation for warnings requires:

1. A reasonable *learning process* involving the hippocampus and basic ganglia, then the prefrontal associative networks.
2. Creating sufficiently strong *long-term memory* so that danger signals will be recognized despite some fade-out or forgetting before recharging.
3. Having a sensitization and *readiness to respond* forged by the early repetition of warning information and subsequent consolidation of relevant knowledge.
4. Facilitating avoidance actions that have been and will be preceded by a likelihood *assessment of the risk* for a given situation and the formation of tentative choices of appropriate alternative avoidance behavior.
5. Ensuring accurate prediction of harm sufficient to *trigger responses* that are based on preplanned logic rather than the sudden initiation of unpredictable, error-prone primitive escape actions.

Such a deliberate and informed process should produce results that are very different from the warnings that result from simplistic, uninformed, quick, and untested efforts expended as almost an afterthought or delegated to persons unsophisticated in the warning specialty. The patient's perspective is that warnings are important to them, and they should be created with the utmost care and social responsibility. If there is an adequate warning analysis, then it can be a good foundation for 1-800 telephone information responders. The information derived from a warning analysis can be applied to other instruction, safety alert, testing, and recall projects.

Pill Matching

To help eliminate wrong-medication and wrong-dose errors during at-home patient self-medication, the patient should have some self-check procedure. One simple and noncostly approach is to print a picture of the pill on the container. The patient can then match the pills in the container to the picture on the container. The container picture should be the same as the picture in the *Physicians Desk Reference* as supplied by the manufacturer for product identification. The pill may have printed information on it, such as RD211, Duricef 500 mg, Hivid 0.750, or Depar 500 mg. This could be better identification than simply pill size, shape, or color. The patient could first check the container pill picture and then check the prescription order from the treating physician. This double-check procedure should substantially reduce patient errors and help to find supply chain errors. It could be of help in the health care facility distribution of drugs to patients.

Instructions

The drug manufacturers provide detailed information about drugs, which is compiled in the annual *Physician's Desk Reference* and semiannual supplements, so that physicians can have the information necessary for intelligent and informed decision making. This compilation is reviewed for accuracy by the medical representatives of the pharmaceutical manufacturers. This drug information, in the same language and emphasis, is supplied to the end user or patient in the form of a package insert. The insert is provided to each patient when the pharmacist fills a prescription and gives it to the patient.

There are several limitations of this information that is intended to supplement the drug labels. First, the insert is in medical language that is far beyond the comprehension of the typical patient. Second, it is rarely read, even in part, by the average patient. Third, it imposes a burdensome requirement for medical education on patients who want to know the risks and benefits of the drug. Fourth, it is unattractive, uninteresting, and in a format that does not provide user-friendly knowledge. Fifth, the technical information can be easily misunderstood by patients of average education, experience, and familiarity with pharmaceutical concepts. Sixth, it is often transmitted in English to foreign-speaking patients. Seventh, the insert sheet is often cut off at the bottom in the attempt to have more readable larger print on a single page. Eighth, a product description that must be ignored creates uncertainty and anxiety, the precursors of medical drug error.

The general insert format includes drug name and common brand names, warnings, uses, how to use, side effects, cautions, drug interactions, overdose information, notes, missed dose instructions, and storage needs. Some pharmacies provide patient consultation information on drugs in more easily understood language. Some surgeons provide their own home care instructions, but the problem of fully, properly, and effectively preventing drug misuse or error persists.

The FDA has issued a rule (docket number 2000N-1269, 2006) describing the format for prescription drug package inserts. For trade labeling, a font size of 6 points is specified for use on the package from which the drug is dispensed. For promotional materials, an 8-point font is specified as a minimum. For prescribing information, a 6- or 7-point font has been used. This is still considered by many to be small print, particularly for elderly patients. A highlights section should provide risk and benefit information, but all the safety information need not be included. How the new rule improves readability remains to be seen. What level of enforcement might occur after the transition process is uncertain. The clear inference is that it remains the prime burden of the prescribing physician to explain the risks and benefits of medications to their patients, a task rarely accomplished.

Regulation

Special Conditions of Use

The U.S. FDA has been given wide regulatory authority over drugs and medical devices to ensure that medical products are both safe and effective. A drug should undergo three stages of clinical trials before approval for use against a particular disease. The general public may believe that the approval process is an assurance that all drugs prescribed are carefully monitored by an appropriate federal regulatory agency. However, the following may occur:

Off-label use can occur because a physician can prescribe a medication approved for one limited use for any use (including those not FDA approved) not included in drug labeling or not listed in package inserts. There is no limitation on the manner in which a physician can use or prescribe a medication in treatment regimens. However, this off-label use should be within acceptable professional practice and not exercised if no relevant safety data exist.

Experimental drugs may be used for seriously ill patients who have exhausted all other commercially available treatments. Patients with cardiovascular diseases, cancer, and human immunodeficiency virus (HIV) have access to drugs still under development and drugs still in clinical trials. Drugs are also available for those who do not fit the criteria for participating in clinical trials or when a drug's development has ceased because it has failed to work in most patients (Dooren 2006c). This may be known as an expanded access program.

There have been continued efforts to broaden access to unproven treatments and for the use of experimental drugs before clinical trials have been concluded. There have been attempts to gain an initial approval, before full approval, so that drug manufacturers can collect data and earn money to support the ongoing drug development process. Others view such use of experimental drugs as an erosion of the FDA mandate, and that there should be good reason to believe a drug will work without subjecting patients to unknown risks.

Antibiotic Approvals

There have been serious questions regarding the FDA approval of antibiotics despite the need for new drug options against drug-resistant bacteria. The drug effectiveness criterion is that new antibiotics should not be inferior to existing drugs. Critics say that new drugs should be proven superior to a placebo. An example of the problem was in the proposed extension of a drug approved for skin infections to use for bacteremia (bacteria in the blood) and endocarditis (heart valve inflammation caused by infections of *Staphylococcus aureus*). Human clinical trial testing of the new drug on infected right heart valves resulted in a success rate of 42% compared with a 44% success rate with vancomycin or a combination of other antibiotics (Mathews and Dooren 2006). Despite the negative opinions of the FDA staff, they were overruled by FDA officials, and approval was given for use against bacteremia and right-side endocarditis. The left valve success was in one of nine compared with two of nine for the older antibiotics. The new drug was considered equivalent to the older drugs by the FDA. The difficulty in treating staph infections was considered since more than half of the staph bacteria found in hospitals were resistant to methicillin, penicillin, and amoxicillin. In general, drug manufacturers want loose standards because clinical trials could cost more than $100,000 per patient, and there is a great need for new drugs. The FDA staff wanted a strongly worded label "warning" about the risk of developing a strain of bacteria that was less susceptible to the drug, but FDA officials decided on a less-potent "precaution" as sufficient to communicate the risks. The justification for the approval of the drug for right-sided endocarditis was that untreated endocarditis is almost always fatal, and that doctors were already using the drug without FDA approval, but at too low a dose (a medical error corrected).

Expert Panels

The FDA advisory committees provide independent advice, but a declining minority of drugs faces advisory committees. There is a high (28%) inconsistency between advisory committee recommendations and subsequent FDA actions. It has been concluded that the FDA makes suboptimum use of its drug advisory committee system (Tapley et al. 2007).

The drug telithromycin is an antibiotic used to treat chronic bronchitis, acute bacterial sinusitis, and community-acquired pneumonia. There were postapproval reports of liver injury, failure, and death associated with its use. Warnings about possible liver injury were then added to the label by the manufacturer. The FDA subsequently sought advice from an expert panel regarding the safety of the drug. An advisory panel then conducted another risk-benefit profile of the drug. At the same time, allegations of fraud in the earlier clinical trials were to be investigated by the Senate Finance Committee and the House Energy and Commerce Committee ("FDA Will" 2006). This scenario illustrates the FDA's reliance on outside expert panels that may or may not include individuals with some possible conflict of interest. What would a patient do, if anything, should the patient notice that a drug label has been changed to include a warning about possible liver damage? Would the patient assume that a warning is routine, the risk is minor, and it is not applicable unless further advised by his or her physician? What information content is actually conveyed to the patients with average or below-average knowledge? It may seem that the de facto influences of an expert panel may be greater than that of the staff of the FDA.

Over-the-Counter Drugs

The FDA cautioned those who take over-the-counter aspirin tablets that there is a risk of gastrointestinal bleeding and kidney injury even when taken at the correct dosage (Bridges 2006). Aspirin is commonly used to treat pain, headache, and fever. The FDA-proposed drug labels highlight the warnings of stomach bleeding for patients older than 59, for those who have stomach ulcers, for those who take blood-thinning drugs, or for those who use drugs containing nonsteroidal anti-inflammatory ingredients. Warnings on painkiller drugs were already extensive, as may be seen on typical aspirin packages, such as those containing 500 tablets. The label print is small, and few consumers read and remember all the various items called warnings.

The FDA indicated that over-the-counter painkillers are safe when used as directed. The intent is to standardize warnings so that all nonprescription packages carry warnings in the same language. The warnings would be more prominent using fluorescent or bold-face type. For acetaminophen, there would be a warning about liver damage. For ibuprofen, aspirin, and painkillers other than acetaminophen, there would be a warning about stomach bleeding.

The consumer medical errors that might be reduced include taking more than the recommended dose, taking the drugs while consuming three or more alcoholic drinks per day, taking multiple drugs containing acetaminophen, and failure to recognize side effects. There should be a search for label expansion, such as the fold-outs used for other consumer products, so that the label warning could have larger-size print and be more readable. The labels should be tested for readability, comprehension, and consumer friendliness among the millions who take such drugs each day.

FDA Limitations

An Institute of Medicine report indicated that the FDA drug approval efforts suffer from organizational and cultural problems, including a scarcity of postapproval data concerning drug risks ("Institute of Medicine Report" 2006). The recommendations indicated that drug labels should clearly denote a product as new for 2 years, with an advertising ban for that 2-year period. The FDA should gather and analyze postapproval data about drug risks, with a formal review of a drug's risks and benefits every 5 years.

Such proposed limitations by a responsible government agency impose a burden on the patients using drugs. What should they believe if a drug has a label marked "new drug"? Is new better than old, or is new really "unproven"? Should the patient be advised to wait for 2 years to see what possible health risks become manifest? Should they wait until the formal risk-benefit review has been completed? There are also questions about advertising bans. Does this include detail, sales, or representative personnel contacting treating physicians and leaving behind descriptive brochures about a new drug? Could it mean no magazine or television advertising aimed directly at the consumer? The pharmaceutical manufacturers may lobby and provide expert panel guidance, but patients rarely have a mechanism by which they can listen or be heard at federal agencies.

There may be postapproval study restrictions. An extensive FDA postapproval study of the controversial silicone augmentation breast implants was approved in November 2006. There had been a 14-year moratorium on silicone or silicone-gel implants because of injury claims and complaints, leaving only the less-desirable and less-costly saline implants on the market. The 10-year postapproval study of 40,000 enrolled or registered women included a provision that defective implants will be replaced free of charge, and that some of the costs of a repeat surgery will be paid by the manufacturer. The cost of the study to one company has been estimated at $45 million and for a second company at more than $30 million. It was estimated that 300,000 women each year would elect to have silicone implants inserted in their bodies, plus additional women around the world. The FDA recommended that implanted women undergo magnetic resonance imaging screening 3 years postimplantation and every 2 years thereafter to monitor for possible silent ruptures and leakage. In a risk-benefit evaluation, how is the benefit (a cosmetic breast augmentation) balanced against the risks (sufficient for a prior moratorium unless there were new design improvements)? Is there medical error involved in this type of postapproval testing on human subjects?

Public Policy Concerns

A report (Whitehead 2006) was critical of the FDA. It reported allegations of a senior drug safety researcher at the FDA that "200,000-plus people are dying every year from prescription drugs." It also repeated the 2005 Center for Public Integrity report stating "the pharmaceutical and health products industry has spent more than $800 million in federal lobbying and campaign donations at the federal and state levels in the past several years." It concluded that "no other industry has spent more money to sway public policy during that period."

Such criticisms reflect the stresses imposed on the FDA by various interest groups, including drug manufacturers, congressional committees, insurers that pay for brand name and generic drugs, academic and pharmaceutical research specialists attempting to advance the frontiers of science, and other advocacy groups.

Prescription Directions

The containers for prescription drugs vary dramatically in the information provided to the patient, user, or consumer. The directions and warnings are usually applied to the container by pharmacists using adhesive-backed message stickers or adhesively mounted vertical tabs. These add-on supplements to the label are typically about 9 mm in height and 40 mm in length (i.e., about ⅜ × 1½ inches). The message font size is about 7 points or smaller. Thus, the actual messages are in fine print that may not be readable by the visually impaired.

These label supplements may have colored backgrounds. Red apparently connotes danger. Yellow suggests caution. Green infers advisory information. The red may be dark, inhibiting the reading of the small black type used for the message. Portions of the message may be in capital letters. There are reports of the use of black triangles for new drugs, black box warnings for high risks, and highlighted messages. Some examples follow:

> *Metformin* (Glucophage): One container has four yellow messages. Another container has five messages: one red, one green, and three yellow. The red sticker has two signal words (warning on top, caution on the side). The message is, "Do not use if you are breast feeding. Consult your pharmacist." Is the message a warning or a caution? It is on only one of the containers, which has no nonlanguage graphics but has two signal words. There are yellow stickers on both containers with the signal word IMPORTANT (in caps with a special black background for the word). Does this mean the other messages are not important? The message is, "Take or use exactly as directed. *DO NOT* discontinue or skip doses unless directed by your doctor." Is this worthy of a warning, a caution, or an advisory signal word? The graphic

is a finger pointed at a label. Both containers have the yellow message *"DO NOT* DRINK ALCOHOLIC BEVERAGES when taking this medication." A crossed (international do not indicator) graphic of a cocktail glass is on the sticker. Both containers have the message "TAKE MEDICATION *WITH FOOD OR MILK."* The graphic is a slice of bread falling from a loaf of bread, which might suggest to some to avoid bread or food. One container has a green sticker located on the cap. There is no signal word. The graphic shows a pill and a capsule. The message reads, "This is the same medication you have been getting. Color, size, or shape may appear different." This tells the consumer that there is no way to verify whether it is the correct medication (see the Pill Matching section in this chapter). Why is this label green as contrasted with other colors used for the label supplements?

Amoxicillin-Clavulanic Acid (Augmention): This container had no supplemental colored stickers. The only message was typed directly on the label as follows: "Take 1 tablet by mouth every 8 hours with or after a meal. Report to us if you have any gastrointestinal upset." This seems to be directions for use, with no warnings.

Glipizide: One container had two supplemental stickers; the other had five stickers. Both had alcohol and before meals stickers; one had a yellow IMPORTANT—use as directed sticker, a red WARNING pregnancy or breast feeding sticker, and a yellow (no signal word) avoid sunlight sticker.

Lisinopril (Zestril): This container had three supplemental stickers. One was a red sticker with no signal word, the message "May cause dizziness," and a graphic of a human eye. The second was a yellow sticker with a signal word "Caution," no graphic, and the message "This medicine may be taken with or without food." Why is this message worth a caution sticker? The third sticker had the signal word "Important" and the message to use as directed.

Levaquin (Levofloxacin): This container had five supplemental stickers. The first was red with the two signal supplement stickers. The first sticker was red with the two signal words Caution and Warning and no graphic. The second was yellow with the signal word Caution and no graphic. The third had a graphic of a faucet filling a glass of water, no signal word, and the message "This medication should be taken with plenty of water." How much water is plenty? The fourth was a yellow dizziness sticker. The fifth had a sunlight avoidance message with a graphic of a crossed gear or sun.

This summary of a short survey clearly indicated problems for the consumer because of nonstandardized supplemental container stickers. They all had small print, which could be easily corrected by the typical fold-out directions and warnings found on other products. There was nonuniformity in the use of color, signal words, and nonlanguage graphics. This important information should be more effectively communicated to those who consume the drugs.

Each prescription drug was accompanied by a Patient Counseling Information sheet that would not be read or comprehended by most people or by a generalized Personal Prescription Information sheet that was slightly more understandable. Some pharmacies had Patient Consultation Information sheets available. Some surgeons had their own Home Care instruction sheets. In essence, there should be research and testing of label directions and warnings to achieve the kind of communication to patients that can serve to avoid medical error.

Security and Counterfeiting

Drug security has been discussed throughout this book because it has been and is a threat to the consumer. Tampering has been costly in the past. A blister pack may be tamper evident if there is good lid seal integrity and a design that immediately shows obvious tears, punctures, intrusions, separations, partial use, or other evidence of product adulteration. There may be the addition of impurities, such as finding the presence of a toxic substance. The drug container may show that it has been opened (a tamper-evident design) or is difficult to open without leaving signs that the integrity of the container has been violated (a tamper-resistant design). Visual inspection for tampering must still occur for patient safety.

Counterfeit pharmaceuticals are a threat to the manufacturer's brand security or identity if they involve a drug sold under a product name without proper authorization. That is, the label suggests an authentic product, but it is intentionally mislabeled regarding the drug identity and source. The counterfeit drugs may contain substituted drugs or incorrect ingredients that cause health problems. The active ingredient may be subpotent or superpotent, resulting in underdosing or overdosing, respectively. The active ingredients may be missing entirely, thus, also posing a consumer threat.

In some countries, counterfeit drugs constitute 10% to 50% of the total market. In the United States, where 4 billion prescriptions were filled in 2005, the counterfeits are difficult to detect because of the sophistication in making copies of legitimate drugs and the volume of drugs in the supply chain. The prescription drug marketplace involves manufacturers, a few primary wholesalers, and some small secondary wholesalers who may buy drugs that do not have a clear linkage back to the manufacturer. Some Internet pharmacies are well known for selling counterfeit drugs. The technical knowledge necessary for counterfeiting has spread to many countries engaged in world trade.

There are a number of remedies to reduce counterfeit drugs. Radio-frequency identification device (RFID) tags provide for an electronic track-and-trace system within the supply chain that yields an accurate drug pedigree (see the Label section of this chapter). The drug manufacturers may include microtaggant identification particles in the drug. These microscopic taggants may have 10 color layers, each assigned a numeric value, so that the color code fingerprint or signature can be read by x-ray analysis or by a photospectrometer (Le Pree 2006). There are also color-shifting inks and holograms for labels and containers.

There may be other causes for variation in the components of a drug, such as substitution of ingredients at the manufacturer's or vendor's facility. These substitutions are based on ingredient availability, improvements in the drug formula, patent implications, or changes in the manufacturing process based on time and cost.

Hospitals should also be security conscious in terms of access to various facilities. In addition to video surveillance, there is the ability to identify each person by fingerprint, facial recognition, handprint, or iris recognition scan (bioidentification). At the elemental level, a worker's identification credential (ID card) may be issued to workers to expedite and simplify the process of entering premises. It may be used to deny access to complex equipment that may have expensive calibration maintenance requirements and is kept in a controlled area. If a no-contact system of identification is needed, then the iris recognition biometric technology can be used. An individual simply looks into the iris reader, and a comparison is then made with the stored image patterns. The recognition time is about 1 second, and the inaccuracy can be less than one error per million. It can be a two-camera, two-iris, no physical contact system that is nonintimidating, portable, and tamperproof. The system may permit tracking of an individual within the premises. Eventually, it could be connected to compatible national databases. Access errors can be virtually nonexistent as compared with human guard and video system monitoring.

Recalls

After successful marketing, some drugs have unexpectedly manifested serious side effects and have been withdrawn from the market. For example, for one drug the *risks* of increased heart attacks and strokes were compared to the *benefits* of painkilling effects, then contrasted to that which might have been achieved by other safer alternatives. The risks outweighed the benefits, and the drug was withdrawn or recalled. Those patients who have been harmed by the recalled drug have some limited remedies. For example, one manufacturer withdrew a drug from the market in 2004 and by 2006 faced 27,200 product liability lawsuits (Tesoriero 2006). Some of the cases had a jury verdict in the millions of dollars and expensive legal costs. From the perspective of the patient, user, or consumer, any adverse event trend should invite severe implications by the printed and video media, induce state attorney general investigations, result in damaging FDA actions, and perhaps result in congressional hearings. These actions may please consumers who become angry at drug failures, but they can seriously damage the reputation of legally and morally responsible pharmaceutical enterprises.

If the costs of defective drugs are so high, then company stakeholders may ask how can such errors be prevented. The consumer may suggest more and better evidence of drug efficacy and safety. There could be greater reliance on strict clinical trials rather than some time- and cost-reducing exceptions to good scientific practice. More and better testing of medical devices could be easily achieved by proper standards compliance, higher levels of design analysis, and better enforcement of good manufacturing practices. The introduction of safety factors in risk-benefit analyses would be a significant consumer advantage. Off-label use and questionable informed consent procedures should be minimized. The incentive of high sales and the burden of high drug development costs may have created another risk-benefit criterion, namely, the seller's benefits of high-dollar sales and profits (seller's monetary advantages) versus the risks assumed by the buyer, which may or may not occur in a fashion to materially harm the seller (seller's disadvantages). There are those who question the very use of risk-benefit analyses in drug and device evaluations and approvals as not sufficiently consumer oriented.

One good solution, from the consumer's viewpoint, may be that the pharmaceutical or device manufacturer should institute an effective and integrated recall planning program. This entails the capture of information on customer dissatisfaction, complaints, claims, rejects, deficiencies, and discrepancies. Next comes an intensive and incisive probe for correctable causes of product problems, possible trends, and human errors with insightful negotiation and cooperation between all company functions. Provision should be made for the institution of real solutions that otherwise might be overlooked, ignored, or subject to quick dismissal as relative to problems that are unimportant or not proven. Attention should be given to why, when, and how recalls can be conducted in the quickest, least-expensive, and most favorable public relations manner. There should be a periodic exercise of simulated recalls to ensure proficiency. There should be recognition that almost all recalls involve human error by specific persons, at given locations, and at a particular time. It is important, in a real recall, that the customer be personally contacted, relieved of any propensity for personal harm, shown personal concern, be emotionally appeased, and be helped to create favorable opinion of the company. Recalls are the prime opportunity to limit liability, prevent harm to the past and prospective end purchaser, and foster good public relations. An early warning or effective recall might save both lives and money.

There have been so many drug recalls and drug withdrawals from the marketplace that some critics allege that drug safety problems are not discovered in the way in which preapproval clinical trials are conducted and evaluated. Similarly, they also claim that recalls are not used quickly enough and sufficiently to properly protect the public from appreciable harm. A case in point occurred in 2007 when two drugs used to treat Parkinson's disease were found to sharply increase the risk of heart valve damage. The drugs stimulate receptors in the brain that mimic the effects of dopamine since Parkinson's disease involves a loss of dopamine, which then causes muscle tremors. One drug manufacturer had issued a safety alert in 2003 about the heart valve side effect. The question was why there was a 3-year delay even though both drug manufacturers "no longer promote the drug" (Hechinger 2007). A study in the United Kingdom of one of the drugs found 19% of the patients suffered new heart valve problems, an increased risk five to seven times that of patients who did not take the drug. Another study, in Italy, found 23% of patients taking one drug and 29% of those taking the other drug suffered heart valve problems compared to a 5.6% rate in a control group. These sharp increases resulted in suggestions to physicians to discontinue the drugs and to have the patients get echocardiograms to evaluate possible heart damage. If patients were to continue with the drugs, then they should be closely monitored. As of January 2007, the drugs were not withdrawn from the market, and no recall had been instituted.

This example illustrates possible balancing of interests in such cases. A recall would not be in the drug manufacturer's financial interests because it might trigger a flood of personal injury lawsuits and adverse publicity. It would be better for the manufacturer to have the drug remain on the market, gradually reduce its sales, caution physicians in order to shift possible blame, and attempt to seek drug alternatives. The FDA interests may favor a voluntary and quiet withdrawal of the drug from the marketplace, avoiding possible conflict with the drug manufacturer if a recall were recommended and not calling attention to its possible failures in the original drug approval and in postapproval surveillance. The treating physician might welcome retaining the choice of drugs, although the burden of close patient monitoring might be unwelcome unless it was ambiguous and voluntary. Mostly overlooked would be the interests of the prospective patients, who might not want to assume the significant risks of a new injury however explained or not explained to them. In such cases, all participants could be jockeying around to transfer the stigma of medical error to other participants.

The importance of decisions regarding drug recall, withdrawal, or limitations in use is illustrated by what happened with an antipsychotic drug that could cause diabetes. There were allegations of a failure to warn patients adequately of the risk. In 2005, the manufacturer paid $700 million to 10,500 product liability claimants. In 2006, there was a settlement of $500 million to 18,000 claimants. In 2007, there were 1,200 holdouts and claims by insurers and state governments (A. Johnson 2007). Aggravating the damages was promotion of the drug for unapproved uses. Reducing the damages was a 2003 addition to the drug label warning of increased diabetic risk. This scenario suggests which factors might be considered in a possible recall when there is a probable ongoing drug problem or medical error. There is no assurance about what may be communicated to the ultimate user, consumer, or patient.

Apparently there was knowledge of a possible problem when the label warning was placed on the product in 2003. Then, there were nearly 30,000 patients claiming injury, $1.2 billion in settlements, and other pending minor and major lawsuits. In 2005 alone, the drug had sales of $4.2 billion. That is huge revenue for a drug, and the liability costs were for cases the company stated were without merit. Should there have been a recall, better warning, restrictions to only approved uses, more shifting of possible liability, or a form of marketing that recognizes the possibility of medical error?

Possible Problems

There may be future problems in the drug delivery system that could beget or spawn medical error.

Complexity

The complexity of manufacturing some pharmaceuticals, which may require more than 100 steps, may involve a difficult and costly series of processes. There may be reduced yields and long cycle times in such situations. The quality control inspections may be impaired, particularly by continual efforts to reduce the number of steps, time, and cost. The actual composition of the drug may vary unpredictably beyond acceptable limits when production changes are made.

Scale-Up

There are commonly known problems moving from laboratory-based or science-based production to online production. There are often similar problems in scaling up for the increased production the market may demand. Such scale-up problems continually surprise both technical and management personnel. The drug output may vary in composition during these events.

Waste

There are always attempts to reduce waste during manufacturing since the potential savings may be in the millions of dollars. Lean may be good, but overzealous reductions may produce errors that could lead to the need for a safety alert to customers.

The pharmaceutical industry in general has inefficient and wasteful manufacturing operations (Mullin 2007). Drug production is usually from a batch-manufacturing operation that reflects the medicinal chemistry drug development process. They are laboratory science small-batch operations. It may be difficult to scale-up to large batches. Yet, there is now a radical shift to continuous processing that involves plant engineering technology, process engineers, process analytical techniques (PATs), Good Manufacturing Practice (GMP) guidelines, and dealing with large-quantity raw material suppliers. To avoid problems and errors, those responsible for process design, development, and management should be involved well before commercial production is scheduled.

Verification

There may not be a third-party quality verification system for active ingredients purchased from worldwide vendors or from producers of outsourced chemical compounds. There may be quality assurance problems with preferred or alternate vendors or from a change to a new vendor. The verification system helps to avoid the use of substandard ingredients and counterfeit materials. No verification means no knowledge or control and the possible marketing of deviant drugs.

Collaboration

For difficult research and development activities, there may be a collaborative agreement between two or more companies. One company may be responsible for discovery of molecules and production of interesting or promising compounds. Another company may focus on preclinical development, clinical evaluation, and commercialization. For example, there may be hundreds of kinase inhibitors and their targets in disease pathways that could be explored to find novel therapeutic agents. There may be a joint search for therapeutic human proteins produced by genetic engineering or DNA (deoxyribonucleic acid) technology that are engineered to be manufactured and purified in high-yield and low-cost processes. Another approach is the acquisition of companies that have made promising advances and could function in a subsidiary corporate role, such as those developing therapeutics based on ribonucleic acid interference (RNAi) or small interfering RNA and their mode of administration to patients. There may be temporary partnerships and other collaborative drug discovery efforts with subsequent breakups and royalty arrangements. Corporate changes often adversely affect discovery, manufacturing, and drug distribution.

Networking and Privacy Protection

Online Searches

Many patients perform information searches on the Internet. They seek personalized health care knowledge. They may attempt to locate support groups that could aid in the coping process. They join e-mail chat rooms and discussion groups to obtain the latest news, for reinforcement of what they believe, and to fill any health care voids that may be important to them. They exchange information on the latest aspects of their disease and suggested or recommended new treatments. They engage in sharing personal information, give advice, help to identify treatment centers, recommend good doctors, and attempt to continuously update what might be beneficial to them.

In the process of sharing information, they may not consider personal privacy rights or possible conflicts of interest. They may use the Web just so they feel that they are not alone or perhaps misinterpreting what they have been told by their caregivers. They obtain descriptive booklets, illustrations, and written discussions about particular diseases and wellness concepts. They may join password-protected sites, visit health fair events, and establish personal relationships. The Web is used not only by patients, but also by many other individuals, fund-raising groups, lawyers, financial advisors, and health care insurance agents. Blogging and message board posting are not restricted to patients. The information may be of great value for others in assessing an individual's future health care options and costs, insurance coverage and premiums, financial goals and models for investment and securities, and fund-raising activities by charitable and for-profit groups. It should be obvious that there are potential privacy and conflict-of-interest problems.

The Internet activities have great value in educating special disease groups and informing them of new treatments or health dangers. They could serve to significantly reduce medical errors and provide for better patient participation in the treatment process. The amount, scope, and detail of the information on the Web is astonishing, in terms of both good and bad, because there is open access and little in terms of expert or peer reviews. The contents of medical journals may be available, but free access is generally limited to out-of-date information. Social networking is rapidly becoming a type of free and open medium for easily understood interactive health information.

Privacy Rules

The Privacy Rule of the 1996 federal Health Insurance Portability and Accountability Act (HIPAA) gives patients access to their medical records and restricts how health care providers can use the records unless there is written consent from the patient. There are also state privacy requirements. A prime objective of HIPAA was to make it easier for people to switch or retain their heath insurance. The carefully constructed privacy rules, present and future, may inadvertently be waived by a patient or be innocently subject to broad and comprehensive written consent forms.

Waiver of Privacy

In China, HIV testing has been associated with stigma and discrimination. The United Nations and the Chinese Ministry of Health estimated the number of people infected with HIV in the People's Republic of China was about 840,000, of whom 80,000 had acquired immunodeficiency syndrome (AIDS) (State Council 2004). Left unchecked, China could have 10 million infected by 2010 (Wu et al. 2006). Control of the epidemic could be hindered by complacency and low participation in voluntary counseling and testing. A national social marketing campaign was instituted to promote HIV awareness and overcome misconceptions. The program included slogans on posters and banners, newspaper and television comments, public announcements by celebrities, and community events. This social marketing approach was based on an infectious disease model. Some question it regarding the protection of human rights since there may be a violation of privacy rights, it may not be entirely voluntary because of social pressure, it may involve little counseling, and it may not involve proper informed consent because it may be done as part of an overall health examination to "normalize" the consent process.

There have been testing initiatives in which provincial governments in China have invited at-risk groups to be tested. These groups include drug users, spouses of HIV-infected individuals, children under 10 years of age whose mothers were HIV positive, patients with sexually transmitted disease, sex workers, former plasma donors, pregnant woman, and patients suffering from infections (Wu et al. 2006). The basic problem of waiving privacy rights varies regarding whether it pertains to one individual in a stable health situation or whether it is necessary to contain an infectious outbreak that could have worldwide ramifications.

Responses

There is considerable value in patient networking. It may result in a better understanding of a disease process, result in improved support and coping procedures, and foster a reduction of worrisome uncertainties. It improves patient involvement in the treatment process. The informed patient should be able to ask the treating physician more relevant questions and better comprehend the responses with less error.

The appropriateness of content in some chat rooms, discussion groups, and Web sites may be questionable or inappropriate. There may be promotion of unproven products and fads, some blatant sales activities, or some illegal conduct, such as fraud and outlawed treatments. The treating physician should inquire about Internet and other networking activity by the patient. The physician may be able to correct misunderstandings that run counter to the prescribed treatment protocol or that could create errors.

In the social networking process, a patient could reveal too much personal information, which might serve as a waiver of privacy rights. There are questions about the ownership of medical records. Is the owner the patient or the medical care specialists? What information could be released to those involved in research or public health endeavors? A networking patient should be cautioned about the release of private information, the effect of informed consent, the key aspects of waivers under the applicable law, and the social benefits of an infectious disease containment program utilizing comprehensive written consent forms.

It may be considered medical error if the patient is not advised about the problems and remedies concerning the inadvertent or unintentional public release of part of a patient's medical record or other relevant personal information that has been given protection under the law. Because of the rapid growth of social networking and the reliance of some patients on its content, this is an area of foreseeable medical error in drug delivery and patient privacy.

Patient Concerns

The patient may feel neglected and overlooked by physicians who require appointments perhaps a week or so in advance. A study indicated that only 40% of primary care physicians in the United States have arrangements for after-hours care, expanded office hours, or care when the office is closed (Mitka 2006b). If there are patient problems with medications after the treating physician's normal office hours, then most patients must wait until the next day to try to arrange an appointment. If the time lag results in an adverse event, then this might be considered a medical error or oversight. It may be compared to the after-hours service rate of 90% in New Zealand, 87% in the United Kingdom, and 76% in Germany.

The patient, in an emergency, may be deprived of personal medical records if the treating physician cannot be contacted. Despite the strong recommendations for eliminating handwritten records with electronic computer systems, in 2006 only 28% of primary care physicians in the United States used electronic health records (Mitka 2006b). This compares with 98% in the Netherlands and 92% in New Zealand.

In Australia, 81% of primary care physicians routinely prescribe medications electronically (Mitka 2006b). This compares to 20% in the United States, despite many recommendations that electronic prescriptions would reduce significant drug delivery medical errors.

Many patients have read about the precautions that can be taken to avoid patient infections. There is a patient fear of hospital-acquired infection since it occurs in 2 million Americans, with 88,000 deaths a year (Rehmann 2007). Patients are aware that small respiratory droplets, contaminated objects, and person-to-person contact could spread antibiotic-resistant strains that are usually not found in public areas. They are aware that clothing of health care personnel may have been splashed or sprayed with another patient's blood, body fluids, secretions, or excretions. They see part-time use of gloves, less than the recommended 15-second hand cleaning, rare use of respirator masks and protective eyeware, other patients nearby that may have highly contagious diseases, and dirty tasks performed in a rather routine manner. Some patients may speak out, but most do not. They need reassurance that there is an adequate infection control program in effect, and that the drug delivery system is carefully monitored to ensure safety.

Patients may have viewed television stories or read newspaper accounts about what has been presented as high infection risks for hospitalized patients. Some may refuse treatment, hospitalization, or medical recommendations because of personal fear or lack of confidence. Even good stories may leave improper impressions. For example, a study was conducted on catheter-related bloodstream infections, which have been depicted as common, costly, and potentially lethal (Pronovost et al. 2006; Seward 2006; Wenzel and Edmond 2006). There are about 48,600 bloodstream infections from central venous catheters in patients who are in intensive care units each year in the United States. The source of infection may be from organisms on the hands of health care workers or from the skin of patients. A 66% reduction in infection was achieved by a daily commitment to a culture of safety, ongoing surveillance by trained infection-control personnel, and a supportive central education program. It included appropriate hand hygiene, use of chlorhexidine for skin preparation, use of full-barrier precautions during the insertion of central venous catheters, use of the subclavian vein as the preferred site for insertion of the catheter, and removal of unnecessary central venous catheters (Wenzel and Edmond 2006).

Another study, published in December 2006, was reported in an Associated Press article that appeared in many newspapers. The headline was "If the 1918 Flu Struck Today, the Death Toll Could Hit 81 Million" (Associated Press 2006). This worldwide estimate was compared with the 1957 flu pandemic that killed 2 million people and the 1968 flu pandemic that killed 1 million people. This and other publications in effect stressed the need for infection control but were based on an assumption of public fear.

Recipients of research grants and fees from the drug industry are often proud of this indication of their professional accomplishments. Financial relationships between pharmaceutical manufacturers (industry) and members of academic institutional review boards (universities) are common (Campbell et al. 2006). This relationship could develop into a conflict-of-interest situation if there were studies conducted or reviewed by the institutional review board member concerning the manufacturer's drugs. This includes drug selection and administration for various diseases, pharmacodynamics, pharmacokinetics, drug actions, drug interactions, drug abuse, adverse reactions, response to drugs, and drug distribution. Some studies have shown that patients have few concerns about the financial ties between their treating physician and companies with drugs under testing. However, the attitude of patients can rapidly change if they are disappointed or have an adverse result. In other words, a favorable attitude or lack of patient concern about a poorly understood relationship can suddenly become a serious patient concern. This form of medical error can be avoided simply by avoiding the conflict of interest. A full disclosure may not be sufficient.

Patients may become concerned if their treating physician gives priority in scheduling to drug salespeople who patients believe give physicians free samples, meals, promotional billboard items, gifts, and other incentives to prescribe the more costly medications. The patient may not understand that some disciplines, such as family medicine, are a complex, low-margin business with high overhead and huge personal time commitments (Brewer 2007). Patient concerns should be recognized and mitigated.

There is much to be gained in patient confidence if the successes, precautions, and solicited patient involvement were better publicized. Uncertainty, fear, or nondisclosure should not prevail. Everyone of interest should participate in medical error prevention during diagnosis, treatment, and drug delivery.

Caveats

Messages

If they do not communicate the desired information in a manner to capture attention and be fully comprehended, then messages are ineffective and useless.

Symbols and Graphics

If tested to ensure effectiveness and the correct interpretation by the target audience, symbols and graphics can convey effective nonlanguage messages and act as reminders and supplements.

Warnings

If warnings do not communicate directly to the user or patient, then they are not warnings at all. If directed at learned intermediaries, then there is no assurance regarding what may be communicated to the ultimate user, consumer, or patient.

Over-the-Counter Drugs

If over-the-counter drugs have been used for decades and if the general public holds beliefs that they are effective and safe based on their experience and that of others, then these drugs may need or require some sort of special label warnings that limit their use if adverse events are discovered.

Labels

Adequate instructions, warnings, and descriptive content should be provided by innovative labels, tags, and fold-outs attached to the container. Supplemental RFIDs should be incorporated into labeling to provide detailed information and to reduce patient confusion and avoidable medical error on the part of the patient and others in the supply chain.

Packaging

It is important that packaging be designed to substantially reduce consumer problems such as unintentional overdosing or underdosing. It should provide needed detailed information in a clear and correct manner. It should provide traceability both in the supply chain and at the point of consumption.

Outserts

Additional information directed to the consumer may require larger-format inserts, with increased copy space, attached to the medication package. This is particularly true when graphic or user-friendly directions and warnings accompany detailed information or multilanguage formats are desirable. The outserts may be up to about a 17 × 39½ inch (43 × 100 cm) flat size, folded to a 120-panel outsert, with each panel 5½ × 1¹³⁄₁₆ inches (14 × 4.6 cm) in size. Some packaging machines, including folding devices, produce 130 panels folded to a 0.375-inch (~1-cm) thickness and attached to one side of a medication package, with each panel 1¼ × 1¼ inches (3 × 3 cm) in size. Such expanded-content labels may be easily resealable, can keep the information where the patient can use it, and can eliminate fine print that is difficult for older patients to read. Other alternates are simple multipanel peel-apart labels, larger glued booklets, or a wraparound label attached to a base label. There is no real excuse for omitting human factors-tailored information that accompanies and is directly attached to the medication package. It is low cost and can substantially reduce medical error.

Internet Prescriptions

The consumer may find health information on the Internet that might be good or profoundly wrong for that individual. The treating physician could write a supplemental Internet prescription that provides Web addresses to acceptable sites. This might help to explain a disease at far greater length than could a physician short on time. Repetition, printouts, and updated information are more easily and timely conveyed to the patient via the Internet. The vast potential of the Internet also includes premium (paid subscription) health services, the digital storage of medical records, and coaching on health insurance claims information.

Patients

Those using drugs at home should be able to match the pills or substances in the container with a picture of the pills on the container to avoid wrong-pill errors by the patient or from the supply chain. The photographs or images of the pill and its markings and the container identification should be the same as that used by the manufacturer for product identification in the *Physicians Desk Reference*. Pills should be marked with brand and item identification.

Disclosure of Errors

It is in the consumer's best interests for each state to enact laws providing evidentiary protection for the immediate disclosure of and for an apology for a medical error. This ensures a full accounting of what went wrong, a personal apology by those responsible to the patient and immediate family members, and an indication of what would be done to prevent the same error from occurring again. The patient should be provided a copy of all relevant medical records and charts as a demonstration of credibility or truthfulness. If appropriate, a reasonable financial offer should be made promptly. Such an open communications approach would reduce the emotional distress to patients and their possible need for reprisals. It would reduce lawsuits as compared to a defend-and-deny and wall-of-silence policy.

Standards

In an age of increasing complexity, multiplying authorities, and more forced choices, there is greater uncertainty regarding appropriate management decisions and forms of operation. Standards provide an acceptable buffer, defense, and indicator of acceptable practice. Standards and regulations may seem burdensome, costly, and an unwelcome intrusion into good practices and informed choices. They are becoming more numerous, cited, and relied on. Compliance should be ensured, not evaded.

Robustness

In the analysis of a health care system, it may be helpful to amplify or exaggerate the noise induced by multifactor interactions. This can be accomplished by simulation or direct observations under stress conditions. The strategy is to accelerate conditions likely to provoke medical error, so that remedies can be fashioned before actual errors occur. It is a discovery task, under conditions of stress, intended to discover and correct weaknesses in the system so that a hardened or more robust organization will result. It attempts to remove uncertainty by a model-based type of analysis, thus improving overall health care management.

Health Care Costs

Medicare prescription drug coverage has been called a "massive social experiment" of the privatized market (McFadden 2007). The health care system has been called "grotesquely inefficient," and it needs "active management" of the market if it is to work well. Comparisons have been made of health care costs in the United States ($6,102 per year per person or 17% of the gross domestic product [GDP]) with the costs in Canada ($3,165 per person or 10% of the GDP). There are those calling for a massive reform of the health care system, including a single-payer system (a government-managed and -financed program) to replace financing through insurers, employers, union pension funds, Medicare/Medicaid programs, and patient payments. The current cost trends, treatment outcomes, and medical error rates may force incremental drastic changes in the practice of medicine in both limitations in high-cost treatments and more burdensome administrative procedures. There will be startup problems, confusion, conflict, life expectancy issues, moral questions, legal issues, increased palliative drug prescriptions, and fewer real patient choices. All of this suggests that medical error and patient safety should be of high concern for those who will attempt to change the system. A proactive application of the information in this book may ease the pain of the oncoming transition and transformation process.

Adverse Events

The FDA has a postmarket database called the Adverse Event Reporting System, intended to detect drug-related side effects and hazards. The volume of adverse event reports now exceeds 400,000 per year. However, there has been severe criticism that the computer system used to sift through the reports is inadequate and needs to be upgraded if the safety tracking program is to be fully effective. An earlier attempt to improve the system was not successful. There have been recommendations also to include adverse event reports on medical devices. As it stands, for the immediate future, great caution should be exercised in utilizing information from or based on the Adverse Event Reporting System.

Noncompliance

More patients are managing their own health care by taking prescribed and over-the-counter medications. Patient noncompliance, a failure to take prescribed medications regularly, occurred in 40% of patients in 2004 in the United Kingdom. They "do not take them regularly enough to derive any benefits" (*Pharmaceutical and Medical Packaging News*, February 2007, p. 40). Noncompliance in the United States is responsible for 10% of all hospital admissions. An analysis of the data in the National Electronic Injury Surveillance System—Cooperative Adverse Drug Event Surveillance project, for 2004 and 2005, indicated mistakes in taking medication and supplements accounted for 700,000 trips to the emergency room each year (*Diabetes Forecast*, March 2007, p. 15). Improvement in patient drug compliance should be a priority. These errors may be reduced by better packaging, including insert calendars that could serve as treatment records, time-marked blister packs, and information that promotes compliance with the dosing schedule. Drugs should be easy to use, carry, and integrate into the patient's normal or everyday life activities.

References and Recommended Reading

Agency for Healthcare Research and Quality. *Making Health Care Safer: A Critical Analysis of Patient Safety Practices.* Evidence Report 43. Washington, DC.

Agency for Healthcare Research and Quality. *Medical Teamwork and Patient Safety: The Evidence-Based Relation.* Publication No. 05-0053. Washington, DC: GPO, April 2005.

Agency for Healthcare Research and Quality. *AHRQ's Patient Safety Initiative: Building Foundations, Reducing Risk.* Publication No. 04-RG005. Washington, DC: GPO, March 2004a.

Agency for Healthcare Research and Quality. *Hospital Survey on Patient Safety Culture.* Publication No. 04-0041. Washington, DC: GPO, September 2004b.

Aggarwal, Rajesh and Darzi, Ara. Technical-skills training in the 21st century. *New England Journal of Medicine* 355(25), December 21, 2006, 2695–96.

Allen, Daphne. Blisters: investing in the wallet. *Pharmaceutical and Medical Packaging News* 14(2), February 2006, 34–40.

AMA Report. *Report of the Council on Medical and Judicial Affairs, 2003 Reference Committee on Amendments to the Constitution and Bylaws of the AMA.*

American Psychiatric Association. *Diagnostic and Statistical Manual of Mental Disorders (DSM-W-TR).* 4th ed. Washington, DC: American Psychiatric Association, 2000.

American Society for Testing and Materials. Medical devices and services. In *Annual Book of ASTM Standards.* Vol. 13.01, Section 13. West Conshohocken, PA: ASTM International, 2006.

Anderson, Jennifer. Medication errors study reveals alarming figures. *Ergonomics Today,* July 24, 2006. (Based on the July 2006 Institute of Medicine report *Preventing Medication Errors.*)

Andrews, John D. and Moss, T. Robert. *Reliability and Risk Assessment.* 2nd ed. New York: American Society of Mechanical Engineers (ASME) Press, 2002.

Armstrong, David. Medical journal editor Nemeroff steps down over undisclosed ties. *Wall Street Journal,* August 28, 2006, B7.

Arnaud, Celia. Portable MRI. *Chemical and Engineering News,* August 8, 2006a, 84(32) 18.

Arnaud, Celia. Driving a spike into viruses. *Chemical and Engineering News* 84(47), November 20, 2006b, 17.

Associated Press. If the 1918 flu struck today, death toll could hit 81 million. *Wall Street Journal,* December 22, 2006, B4. (Based on an article by Chris Murray in *Lancet,* December 2006.)

Atkinson, Ronald J. Gene regulation of lung cancer. In Peters, George A., and Peters, Barbara J. (Eds.), *Sourcebook on Asbestos Diseases: Pathology, Immunology, and Gene Therapy.* Vol. 17. Charlottesville, VA: Reed Elsevier, 1998, 1–17.

Australian Standard AS 2342, Part 3, 1980 Test Procedures for Evaluating Graphic Symbols and Symbol Signs, 1981.

Ayas, Najib T., Barger, Laura K., Cade, Brian E., et al. Extended duration work and the risk of self-reported percutaneous injuries to interns. *JAMA* 296(9), September 2006, 1055–62.

Aziz, Mohammed Abdel, et al. Epidemiology of antituberculosis drug resistance (the Global Project on Anti-tuberculosis Drug Resistance Surveillance): An update analysis. *Lancet* 368, December 16, 2006, 2142–54.

Baby talk: Drug firm's cash sways debate over test for pregnant women. *Wall Street Journal,* December 13, 2006, A12.

Bachtold, Daniel. Conflict-of-interest allegations derail inquiry into antidepressant's "dark side." *Science* 300, April 4, 2003, 33.

Baddeley, Alan. Working memory. *Science,* January 31, 1992, 556–59.

Baker, David P., Gustafson, Sigrid, Beaubien, Jeff, Salas, Eduardo, and Borach, Paul. *Medical Teamwork and Patient Safety: The Evidence-Based Relation.* Publication No. 05-0053. Rockville, MD: Agency for Healthcare Research and Quality, April 2005.

Baker, G. Ross, Norton, Peter G., et al. The Canadian Adverse Events Study. *Canadian Medical Association Journal* 170(11), 1040–1498, May 25, 2004.

Barnett v. Merck, E.D. La. 2006, No. 2:06-cv-00485-EEF-DEK.

Bazell, Robert. Potentially lethal heart devices a frightening problem for patients. www.msnbc.msn.com/id/15816251/from/ET. Updated November 21, 2006.

Benetez, Yani, Forrester, Leslie, Hurst, Carolyn, and Turpin, Debra. Hospital reduces medication errors using DMAIC and QFD. *Quality Progress,* January 2007, 38–45.

Berkow, Robert, Beers, Mark H., and Fletcher, Andrew J. (Eds.). *The Merck Manual of Medical Information.* Whitehouse Station, NJ: Merck Research Laboratories, 1997.

Bermejo, Magdalena, Rodriguez-Teijeiro, Jose Domingo, Illera, Germain, Barroso, Alex, Vila, Carles, and Walsh, Peter. Ebola outbreak killed 5,000 gorillas. *Science* 314, December 8, 2006, 1564.

Bickley, Lynn S., and Szilagyi, Peter G. *Bates Guide to Physical Examination and History Taking.* 9th ed. Philadelphia: Lippencott Williams and Wilkins, 2007.

Blendon, Robert J., Schoen, Cathy, DesRoches, Catherine, Osborn, Robin, and Zapert, Kinga. Common concerns amid diverse systems: Health care experiences in five countries. *Health Affairs* 22(3), May–June 2003, 106–21.

Bliss, Tim V. P., Collingridge, Graham L., and Laroche, Serge. ZAP and ZIP, a story to forget. *Science* 313, August 25, 2006, 1058–59.

Borwegen, Bill. Airborne infections and respirators. *American Journal of Nursing* 106(10), October 2006, 33–34.

Box, George E. P., Hunter, William G., and Hunter, J. Stuart. *Statistics for Experimenters: An Introduction to Design Data Analysis and Model Building.* 2nd ed. Hoboken, NJ: John Wiley and Sons, 2005.

Brennan, T. P., Leape, L. L., Laird, N. M. et al. Incidence of adverse events and negligence in hospitalized patients: Results from the Harvard Medical Practice Study. *New England Journal of Medicine* 324, 1991, 370–76.

Brewer, Benjamin. I stopped seeing drug sales reps and had more time for patients. *Wall Street Journal,* January 9, 2007, B10.

Bridges, Andrew. Pain-drug warnings get extra strength. *Daily Breeze,* December 20, 2006, pp. 1, A10

Burtka, Allison T. Scrutiny of defibrillator defects grows. *Trial,* August 2006, 68–70.

Campbell, Eric G., et al. Financial relationships between institutional review board members and industry. *New England Journal of Medicine,* 355(22), November 30, 2006, 2321–37.

Carbone, Michele, Powers, Amy, Pass, Harvey, et al. Asbestos, Simian Virus 40, and the development of malignant mesothelioma. In Peters, George A., and Peters, Barbara J. (Eds.), *Sourcebook on Asbestos Diseases: Right to Care, Epidemiology, Virology, Gene Therapy, Legal Defenses.* Vol. 18. Charlottesville, VA: Reed Elsevier, 1998, 269–87.

Casrouge, Armanda, et al. Herpes simplex virus encephalitis in human UNC-93B deficiency. *Science* 314, October 13, 2006, 308–11.

CEN. *Chemical and Engineering News,* August 21, 2006, 36.

Chaffee, Mary. Making the decision to report for work in a disaster. *American Journal of Nursing* 106(91), September 2006, 54–57.

Chang, Alicia. Hospital disease control. *Santa Monica Daily Press,* December 22, 2006, 9.

Chen, Wenhao, Jianshun, Jensen, Zhang, S., and Kosar, Douglas. Air cleaners for removing indoor particles and volatile organic compounds (VOCs). *Engineered Systems,* August 2006, 4–5.

Chin, Gilbert, and Yeston, Jake. Editor's choice. *Science,* 313, July 28, 2006, 411.

Clayton, Margaret F. Communications: An important part of nursing care. *American Journal of Nursing* 106(11), November 2006, 70–72.

Code of Ethics for Nurses with Interpretive Statements. Washington, DC: American Nurses Association, 2001.

Cohen, Michael R. Medication errors. *Nursing 2005* 35(10), October 2005, 10.

Colorful wristbands enhance emergency care. *Nursing 2005,* 33.

Conklin, Joseph D. DOE and Six Sigma. *Quality Progress,* March 2004, 66–69.

Cookson, Mark R., and Hardy, John. The persistence of memory. *New England Journal of Medicine* 355(25), December 21, 2006, 2697–98.

Cooper, Jeffrey B., Newbower, Ronald, Long, Charlene D., and McPeek, Bucknam. Preventable anesthesia mishaps: A study of human factors. *Anesthesiology* 49, 1978, 399–406.

Corina, David P., Vaid, Jyotsna, and Bellugi, Ursula. The linguistic basis of left hemisphere specialization. *Science,* March 6, 1992, 1258–60.

Couzin, Jennifer. Proposed guidelines for emergency research aim to quell confusion. *Science* 313, September 8, 2006, 1372–73.

Crinion, J. et al. Language control in the bilingual brain. *Science* 312, June 9, 2006, 1537–40.

Dagle, Jeff. Using bad data to train good grid dispatchers. *Utility,* September 2006, 26–31.

Daston, Lorraine, and Gigerenzer, Gird. The problem of irrationality. *Science,* June 2, 1989, 1094–95 (book review of Margolis, Howard, *Patterns, Thinking, and Cognition.* Chicago, IL: University of Chicago Press, 1988).

Davis, J. R. (Ed.). *Handbook of Materials for Medical Devices.* Materials Park, OH: American Society of Metals (ASM), 2003.

Davis, P., Lay-Yee, R., Briant, R., Ali, W., Scott, A., and Shug, S. Adverse events in New Zealand public hospitals: Occurrence and impact. *New Zealand Medical Journal* 115(1167), 2000, U271.

Dawood, Richard. Fight the flu. *Safety and Health Practitioner* 24(17), November 2006, 49–50.

DeRenzo, Evan G., Sandler, Alan L., and Moss, Joel. Ethical considerations in the creation of research protocols in pulmonary disease. In Peters, G. A., and Peters, B. J. (Eds.), *Sourcebook on Asbestos Diseases,* Vol. 15. Charlottesville, VA: Lexis, 1997, chap. 1.

Dooren, Jennifer Corbbett. Restrictions curb use of powerful acne drug. *Wall Street Journal,* September 12, 2006a, D1, D4.

Dooren, Jennifer Corbbett. Spinal-disk device is recommended by FDA panel. *Wall Street Journal*, September 20, 2006b, A27.

Dooren, Jennifer Corbett. FDA pushes access to experimental drugs. *Wall Street Journal*, December 12, 2006c, A9.

Draycott, Tim, Sibanda, Thalani, Owen, Louise, et al. Does training in obstetric emergencies improve neonatal outcome? *BJOG An International Journal of Obstetrics and Gynaecology*, 2006, 177–82.

Dutton, Mark. *Orthopaedic Examination, Evaluation, and Intervention*. New York: McGraw-Hill Medical, 2004.

E. coli enablers [editorial]. *Wall Street Journal*, December 18, 2006, A16.

Editorial. Time for a debate on health care in the USA. *Lancet* 368(9540), September 16, 2006, 963.

Editorial. Developing country trialists are key for malaria vaccine goals. *Lancet* 368(9554), December 23, 2006 to January 5, 2007, 2185.

Endsley, Mica R. Toward a theory of situation awareness in dynamic systems. *Human Factors* 37(1), 1988, 32–64.

Enserink, Martin. Test kit error is wake-up call for 50-year-old-foe. *Science* 308, April 22, 2005, 476.

Ericson, Clifton A. *Hazard Analysis Techniques for System Safety*. Hoboken, NJ: Wiley, 2005.

Erlich, Peter F., Rockwell, Sherry, Kincaid, Stephanie, and Mucha, Peter. American College of Surgeons, Committee on Trauma Verification Review: Does it really make a difference? *Journal of Trauma, Injury, Infection, and Critical Care* 53(5), November 2002, 811–16.

Esimai, Grace O. Lean Six Sigma reduces medication errors. *Quality Progress*, April 2005, 51–57.

Ethics and Human Rights Position Statements: Risk versus Responsibility in Providing Nursing Care. Washington, DC: American Nursing Association, 1994.

European Union Regs. *Chemical Information*, July 2006, 27.

Everts, Sarah. Bacterial conversations. *Chemical and Engineering News*, October 23, 2006a, 17–26.

Everts, Sarah. Copper capture. *Chemical and Engineering News*, December 11, 2006b, 12. (Also see *Nature Chemical Biology*, DOI 10.1038/nchembio844).

FDA will review Ketek. *Chemical and Engineering News*, November 20, 2006, 58.

Federal Judiciary Center, *Reference Manual on Scientific Evidence*. 2nd ed. Albany, NY: Matthew Bender, 2000, 333–84.

Financial relationships between institutional revision board members and industry. *New England Journal of Medicine* 355(22), November 30, 2006, 2321–37.

Fisher, Ronald A. *The Design of Experiments*. 9th ed. New York: Hafner Press, 1974 (1st edition published in 1935).

Flin, Rhona, Fletcher, Georgina, McGeorge, Peter, Sutherland, Andrew, and Patney, Rona. Anaesthetists' attitudes to teamwork and safety. *Anaesthesia* 58, 2003, 233–42.

Floto, R. Andres, MacAry, Paul A., et al. Dendritic cell stimulation by mycobacterial Hsp 70 mediated through CCR5. *Science* 314, October 20, 2006, 454–58.

Fontanazza, Maria. Hospitals demand switch to PVC- and DEHP-free devices. *Medical Device and Diagnostic Industry*, February 2006, 22, 24.

Foucar, Elliott. There are ways to ensure valid science trials. *Wall Street Journal*, October 9, 2006, A19.

Freemantle, Michael. Tagging based on molecular logic. *Chemical and Engineering News*, Sepember 11, 2006, 11.

Fuhrmans, Vanessa. Fight over health claims spawn a new arms race. *Wall Street Journal*, February 14, 2007, pp. 1, A18.

Gaba, David M., Howard, Steven K., and Small, Stephen D. Situation awareness in anesthesiology. *Human Factors* 37(1), 1995, 20–31.

Gazzaniga, Michael S. Organization of the human brain. *Science*, September 1, 1989, 947–51.

Ge, Woo-Ping, et al. Long-term potentiation of neuron-glia synapses mediated by CA^{2+}-permeable AMPA receptors. *Science* 312, June 9, 2006, 1533–37.

Gieras, Isabella, and Ebben, Steve. Advancing patient safety through human factors engineering. *Medical Design Technology*, February 2007, 30.

Gish, Gareth B. Medical ventilation. In Young, J. A., and Crocker D. (Eds.), *Principles and Practice of Respiratory Therapy*. 2nd ed. Chicago: Year Book Medical Publishing, 1976, chap. 19.

Gray, James E., Suresh, Gautham, Ursprung, Robert, et al. Patient misidentification in the neonatal intensive care unit: Quantification of risk. *Pediatrics* 117(1), January 2006, 43–47.

Grebow, Matt. Recalls of blood glucose meters emphasize human factors design. *IVD Technology*, January–February 2006, 14–15.

Green, Lois, Melnick, Glenn A., and Nawathe, Amar. *On-Call Physicians at California Emergency Departments*. Oakland, CA: California Health Care Foundation, 2005.

Green-McKenzie, Judith, and Caruso, Garson. Health care workers' crucial barriers. *Occupational Health and Safety*, April 2006, 57–58.

Grissinger, Matthew, and Globus, Nancy J. How technology affects your risk of medication errors. *Nursing 2004* 34(1), January 2004, 36–41.

Gurerk, Ozgur, Irlenbusch, Bernd, and Rockebach, Bettina. The competitive advantage of sanctioning institutions. *Science*, 312, April 7, 2006, 108–10.

Harkin, Tom, Committee on Appropriations, Departments of Labor, Health, and Human Services, and Education, and Related Agencies. *Appropriation Bill, 2002, Report 107-84*. Washington, DC: GPO, October 11, 2001.

Hechinger, John. Heart risk is shown in Parkinson's drugs. *Wall Street Journal*, January 4, 2007, A4.

Hellman, Richard. A systems approach to reducing errors in insulin therapy in the inpatient setting. *Endocrine Practice* 10(Suppl. 2), March–April 2004, 100–108.

Henderson, John. *Emergency Medical Guide*. 3rd ed. New York: McGraw Hill, 1973.

Hendrick, Hal W., and Kleiner, Brian M. (Eds.). *Macroergonomics: Theory, Methods, and Applications*. Mahwah, NJ: Lawrence Erlbaum Associates, 2002.

Henrich, Joseph. Cooperation, punishment, and the evolution of human institutions. *Science* 312, April 7, 2006, 60–61.

Hermans, Michel H. Silver-containing dressings and the need for evidence. *American Journal of Nursing* 106(12), December 2006, 60–67.

Herper, Matthew, and Langreth, Robert. Dangerous devices. *Forbes*, November 27, 2006, 94–98.

Herranz, Gonzalo, and Ruiz-Canela, Miguel. Institutional review boards and informed consent. In Peters, G. A., and Peters, B. J. (Eds.), *Sourcebook on Asbestos Diseases*. Vol. 21. Charlottesville, VA: Matthew Bender/Lexis, 2000, chap. 1.

Hoffman, Catherine B. Simple truths about America's uninsured. *American Journal of Nursing*, January 2007, 40–47.

Hohenhaus, Susan, Powell, Stephen, and Hohenhaus, Jay T. Enhancing patient safety during hands-off. *American Journal of Nursing* 106(81), August 2006, 72A–72C.

Hospital Bed System Dimensional and Assessment Guidance to Reduce Entrapment—Guidance for Industry and FDA Staff. Washington, DC: Center for Devices and Radiological Health, U.S. Food and Drug Administration, 2006.

Ignacio, Joselito S. Effectively assessing exposure to pandemic influenza. *The Synergist* 17(9), October 2006, 36–40.

Institute of Medicine. *Preventing Medication Errors: Quality Chasm Series.* Washington, DC: National Academy Press, 2006.

Institute of Medicine report slams the FDA. *Quality Progress* 39(11), November 2006, 16.

Isbell, L. M., Smith, H. L., and Wyer, R. S. Consequences of attempts to disregard social information. In Golding, Jonathan M., and MacLeod, Colin M. (Eds.), *International Forgetting: Interdisciplinary Approaches.* Hillsdale, NJ: Lawrence Erlbaum Associates, 1998, 289–320.

ISO 7010: 2003 Graphical Symbols—Safety Colours and Safety Signs—Safety Signs Used in Workplaces and Public Areas. 2003.

Jacoby, Marx. EU chemicals proposal prompts global mobilization led by U.S. *Wall Street Journal,* June 27, 2006, A6.

Johnson, Avery. Lilly settles 18,000 Zyprexa claims, but more loom. *Wall Street Journal,* January 5, 2007, A4.

Johnson, J. A., Gitlow, H., Widener, S., and Popovich, E. Designing new housing at the University of Miami: A "Six Sigma" DMADV/DFSS case study. *Quality Engineering* 18(3), September 2006, 299–323.

Joint Commission on Accreditation of Healthcare Organizations. A follow-up review of wrong site surgery. *Sentinel Event Alert,* Issue 24, December 5, 2001.

Joint Commission on Accreditation of Healthcare Organizations. Root causes of medication errors 1995–2003. 2004. http://www.jcaho.org/accredited+organizations/ambulatory+care/sentinel+events/rc+of+medications+errors.htm.

Kannally, Tom. Disposable interconnects enable enhanced treatment options. *ECNmag,* January 2007, 27.

Kayyali, Andrea. Drug-resistant bacteria in hospital patients. *American Journal of Nursing* 106(9), September 2006, 721 (re: Furuno, J. P., et al., *Archives of Internal Medicine* 166(5), 2006, 580–85).

Kepner, Charles A., and Tregoe, Benjamin B. *The Rational Manager: A Systematic Approach to Problem Solving and Decision Making.* New York: McGraw Hill, 1965.

Kira, Alexander. *The Bathroom.* New York: Viking, 1976.

Klein, George. A cellular and molecular foundation for understanding cancer. *Science* 313, August 11, 2006, 762–63.

Knauth, Peter. Workload length and shiftload issues. In Marras, William S., and Karowski, Waldemar (Eds.), *Interventions, Controls, and Applications in Occupational Ergonomics.* 2nd ed. Boca Raton, FL: Taylor and Francis, 2006, chap. 29.

Knoch, Daria, Pascual-Leone, Alvaro, Meyer, Kaspar, Treyer, Valerie, and Fehr, Ernst. Diminishing reciprocal fairness by disrupting the right prefrontal cortex. *Science* 314, November 3, 2006, 829–32.

Kohn, Linda T., Corrigan, Janet M., and Donaldson, Molla S. (Eds.). *To Err Is Human: Building a Safer Health System.* Committee on Quality of Health Care in America, Institute of Medicine. Washington, DC: National Academy Press, November 30, 1999.

Konz, Stephen. Rest allowances. In Marras, William S., and Karwowski, Waldermar (Eds.), *Fundamentals and Assessment Tools for Occupational Ergonomics.* 2nd ed. Boca Raton, FL: Taylor and Francis, 2006, chap. 38.

Lambert, Edward C. *Modern Medical Mistakes.* Bloomington, IN: University Press, 1978.

Landrigan, Christopher P., Barger, Laura K., Cade, Brian E., Ayas, Najibt and Czeisler, Charles. Interns' compliance with Accreditation Council for Graduate Medical Education work-hours limits. *JAMA* 296(9), September 2006, 1063–70.

Landro, Laura. Hospitals take stronger steps against bacteria. *Wall Street Journal*, March 8, 2006a, D1.

Landro, Laura. Hospitals move to cut dangerous lab errors. *Wall Street Journal*, June 14, 2006b, D1.

Landro, Laura. Hospitals combat errors at the "hand off." *Wall Street Journal*, June 28, 2006c, D-1.

Landro, Laura. Preventing the tragedy of misdiagnoses. *Wall Street Journal*, November 29, 2006d, D-1, D-5.

Landro, Laura. Hospitals take consumers' advice. *Wall Street Journal*, February 7, 2007, D-3.

Lazare, Aaron. Apology in medical practice: An emerging clinical skill. *JAMA* 296(11), September 20, 2006, 1401–4.

Leape, Lucien. Error in medicine. In Rosenthal, Marilynn, Mulcahy, Linda, Lloyd-Bostock, Sally (Eds.). *Medical Mishaps*. Buckingham: Open University Press, 1999, 20–38.

Le Pree, Joy. Risky business. *Chemical Information.*, December 2006, 16–19.

Lieberson, Alan D. *Healthcare Enterprise Liability*. Charlottesville, VA: Lexis, 1997 (with Cumulative Supplements).

Lovely, Erica. Caregivers get some relief. *Wall Street Journal*, December 23, 2006, A4.

Machalaba, Daniel. Railroads push to harden tank cars. *Wall Street Journal*, September 18, 2006, B2.

Martin v. Hacker, 83 N.Y.2d 1, 628 N.E.2d 1308 (1993).

Marx, Jean. Puzzling out the pains in the gut. *Science* 315, January 5, 2007, 33–35.

Mason, Malia F., Norton, Michael I., Van Dorn, John D., Wegner, Daniel, M., Grafton, Scott T., and Macrae, C. Neil. Wandering minds: The default network and stimulus-independent thought. *Science* 315, January 19, 2007, 393–95.

Mass spec analysis of bacterial biofilms. *Chemical and Engineering News*, January 8, 2007, 48.

Mathews, Anna Wilde. Drug firms use financial clout to push industry agenda at FDA. *Wall Street Journal*, Sepember 1, 2006a, A10.

Mathews, Anna Wilde. Inside FDA, a battle over drug to treat "Darth Vader" bacteria. *Wall Street Journal*, December 1, 2006b, pp. 1, A10.

Mathews, Anna Wilde, and Dooren, Jennifer Corbett. FDA drops panelist in conflict case. *Wall Street Journal*, September 13, 2006, A11.

Maxwell, David, Grenny, Joseph, McMillan, Ron, Patterson, Kerry, and Switzer, Al. *The Seven Crucial Conversations for Healthcare*. Aliso Viejo, CA: American Association of Critical Care Nurses and VitalSmarts, 2005.

Mayor, Susan. Hospital acquired infections kill 5,000 patients a year in England. *British Medical Journal* 321, 2000, 1370.

McCartney, Scott, Reducing goofs to make flights safer. *The Wall Street Journal*, 15 September 2005, p. D5.

McFadden, Daniel L. A dog's breakfast. *Wall Street Journal*, February 16, 2007, A15.

McGregor, Douglas. *The Human Side of the Enterprise*. New York: McGraw-Hill, 1960.

McKay, Betsy. Disease expert from China is nominated as WHO director. *Wall Street Journal*, November 9, 2006, B7.

McPhillips, Heather A., Stille, Christopher J., Smith, David, et al. Potential medication dosing errors in outpatient pediatrics. *Journal of Pediatrics*, December 2005, 761–67.

McRae, Ronald. *Pocketbook of Orthopaedics and Fractures*. 2nd ed. London: Churchill Livingstone Elsevier, 2006.

McWilliams, Gary. Big employers plan electronic health records. *Wall Street Journal,* November 29, 2006, B1, B3.

Meade, Christine M., Bursell, Amy L., and Ketelsen, L. Effects of nursing rounds. *American Journal of Nursing* 106(9), September 2006, 58–70.

Michaels, Harold T. Anti-microbial characteristics of copper. *ASTM Standardization News,* October 2005, 28–31.

Miller, Greg. Probing the social brain. *Science* 312, May 12, 2006a 838–39.

Miller, Greg. A better view of brain disorders. *Science* 313, September 8, 2006b, 1376–79.

Miller, Greg. A surprising connection between memory and imagination. *Science* 315, January 19, 2007, 312.

Mills, Edward, and Rennie, Stuart. HIV testing and individual rights. *Science* 314, October 20, 2006, 417.

Mishap in weapon dismantling nearly leads to detonation. *Journal of System Safety,* January 2007, 23.

Mitka, Mike. Critics say drug-eluting stints overused. *JAMA* 296(17), November 1, 2006a, 2077.

Mitka, Mike. Electronic health records, after-hours care lag in U.S. primary care practices. *JAMA* 296(24), December 27, 2006b, 2913–14. (Based on an article by Schoen, Cathy, et al., in *Health Affairs,* November 2006, 555–71.)

Mokhberry, Javad. Making sense of orthopedic technologies. *Medical Design Technology,* February 2007, 17.

Mullin, Rick. Breaking down barriers. *Chemical and Engineering News,* January 22, 2007, 11–17.

Mundel, August B. *The Product Recall Planning Guide,* 2nd ed. Milwaukee, WI: American Society for Quality, 1999.

Murphy, Eleanor. Patient care: The team approach. In Young, J. A., and Crocker, D. (Eds.), *Principles and Practice of Respiratory Therapy.* 2nd ed. Chicago: Yearbook Medical Publishers, 1976, chap. 16.

Musculoskeletal Disorders in the Workplace. Washington, DC: National Academy of Sciences, 2001.

Myerburg, Robert J., Feigal, David W., and Lindsay, Bruce. Life-threatening malfunction of implantable cardiac devices. *New England Journal of Medicine* 354(22), June 1, 2006, 2309–11.

Naccache, Lionel. Is she conscious? *Science* 313, September 8, 2006, 1395–96.

Naik, Gautam. To reduce errors, hospitals prescribe innovative designs. *Wall Street Journal,* May 8, 2006, pp. 1, A-16.

Naik, Gautam. Faltering family MDs get technology lifeline. *Wall Street Journal,* February 23, 2007, 1.

Naim, Leon Y., and Lemesch, Curt. Asbestos standards around the world. In Peters, George A., and Peters, Barbara J. (Eds.), *Sourcebook on Asbestos Diseases.* Vol. 8. Salem, NH: Butterworth, 1993, 273–85.

Neergaard, Lauran. Getting to the heart of the issue of transplants. *Daily Breeze,* October 31, 2006, pp. 1, A9.

Nelson, Audrey. Patient handling in health care. In Marris, William S., and Karwowski, Waldemar (Eds.), *Interventions, Controls, and Applications in Occupational Ergonomics.* Boca Raton, FL: Taylor and Francis, 2006, chap. 46.

Nelson, Ellen. Cleaning and sterilization. In Young, J. A., and Crocker, D. (Eds.), *Principles and Practice of Respiratory Therapy.* 2nd ed. Chicago: Year Book Medical Publishers, 1976, chap. 11.

Nimmo, Ian. Effective shift handover is no accident. *Chemical Processing*, June 2006, 30–33.

Normile, Dennis. New H5N1 strain emerges in Southern China. *Science* 314, November 3, 2006, 742.

Normile, Dennis. Indonesia taps village wisdom to fight bird flu. *Science* 315, January 5, 2007, 30–33.

Obid, Ghassan. Strategic care. *Waste Management World*, July–August 2006, 144–49.

Oien, Kathryn M., and Goernert, Phillip N. The role of intentional forgetting in employee selection. *Journal of General Psychology*, 130(1), 2003, 97–110.

Owens, Adrian M., Coleman, Martin R., et al. Detecting awareness in the vegetative state. *Science* 313, September 8, 2006, 1402.

Pai, Sanjay A. Laboratory tests: Proper communication reduces error. *SBMJ* (13) November 2005, 397–440.

Patterson, Patricia A., and North, Robert A. Fitting human factors in the product development process. *Medical Device and Diagnostic Industry*, January 2006, 124–32.

Pellicone, Angelo, and Martocci, Maude. Faster turnaround time. *Quality Progress*, March 2006, 31–36.

Peters, George A. *Human Error Principles*. Report ROM 3181-1001. Canoga Park, CA: Rocketdyne, a division of North American Aviation, Canoga Park, CA, March 1, 1963. (Also in *Missile System Safety: An Evaluation of System Test Data*, Report R-5135 (NASA N 63-15092), Rocketdyne, and reprinted in AFSC *Safety Newsletter*, August–September 1963, pp. 14–16, and *Approach* magazine, May 1964.)

Peters, George A. Human error prevention. In Peters, G. A., and Peters, B. J. (Eds.), *Asbestos Health Risks*. Volume 12 of *Sourcebook on Asbestos Diseases*. Charlottesville, VA: Michie, Division of Reed Elsevier, 1996, 207–34.

Peters, George A. Product liability and safety. In Kreith, F. (Ed.), *The CRC Handbook of Mechanical Engineering*. Boca Raton, FL: CRC Press, 1998, 20-11 to 20-15. (In the second edition, 2005, see pp. 20.2 to 20.16.)

Peters, George A., Hall, Frank, and Mitchell, Charles. *Human Performance in the Atlas Engine Maintenance Area*. ROM 2181-1002. Canoga Park, CA: Rocketdyne, a Division of North American Aviation, 1962.

Peters, George A., and Peters, Barbara J. *Sourcebook on Asbestos Diseases*. New York: Garland STPM Press, 1980, H-29–H-31.

Peters, George A., and Peters, Barbara J. Editors' commentary. The treatment and prevention of asbestos diseases. In *Sourcebook on Asbestos Diseases*. Vol. 15. Charlottesville, VA: Lexis, 1997, 284.

Peters, George A., and Peters, Barbara J. *Warnings, Instructions, and Technical Communications*. Tucson, AZ: Lawyers and Judges Publishing Company, 1999.

Peters, George A., and Peters, Barbara J. Driver distractions: Building in human error. *Rx Law and Medicine Report* 2(4), Summer 2000, 3–4.

Peters, George A., and Peters, Barbara J. *Automotive Vehicle Safety*. London: Taylor and Francis, 2002a.

Peters, George A., and Peters, Barbara J. The expanding scope of system safety. *Journal of System Safety* 38, 2002b, 12–16.

Peters, George A., and Peters, Barbara J. Design safety compromises. *Journal of System Safety* 40(6), 2004, 26–29.

Peters, George A., and Peters, Barbara J. Critical issues for system safety management. *Journal of System Safety*, October 2006a, 17–21, 37.

Peters, George A., and Peters, Barbara J. *Human Error: Causes and Control*. London: CRC Press/Taylor and Francis, 2006b.

Peters, George A., and Peters, Barbara J. Legal issues in occupational ergonomics. In Marras, William S., and Karwowski, Waldemar (Eds.), *Occupational Ergonomics, Fundamentals and Assessment Tools for Occupational Ergonomics*. Boca Raton, FL: CRC Press/Taylor and Francis, 2006c, chap. 3.

Peters, George A., and Peters, Barbara J., The Ten Challenges. *The Synergist*, September 2006d, 33 to 37 and October 2006, 32–33.

Porter, Michael E., and Teisberg, Elizabeth O. Redefining competition in healthcare. *Harvard Business Review*, June 2004, 65–70.

Price, Bernie. Improve operational accuracy. *Chemical Processing*, December 2005, 19–23.

Pronovost, Peter, et al. An intervention to decrease catheter-related bloodstream infections in the ICU. *New England Journal of Medicine* 355(26), December 28, 2006, 2725–32.

Quality Interagency Coordination Task Force. *Doing What Counts for Patient Safety: Federal Actions to Reduce Medical Errors and Their Impact*. Washington, DC: GPO, February 2000.

Raichle, Marcus E. The brain's dark energy. *Science* 314, November 24, 2006, 1249–50.

Reduce patient infection with a touch. *Building Design and Construction*, November 2006, 77.

Rehmann, Terri. The essentials of PPE. *Nursing Made Incredibly Easy* 5(1), January–February 2007, 30–39.

Reid, R. Dan. Developing the voluntary care standard. *Quality Progress* 39(11), November 2006, 68–71.

Reuters News Service. Wyeth loses Prempro trial in Philadelphia. *Wall Street Journal*, January 30, 2007, B13.

Roethlesberger, F. I., and Dickson, W. J. *Management and the Worker*. Cambridge, MA: Harvard University Press, 1939.

Rosenthal, Marilynn M., and Sutcliffe, Kathleen M. (Eds.). *Medical Error: What Do We Know? What Do We Do?* San Francisco: Jossey-Bass, 2002.

Rubenstein, Lynn. Cleaning up medical waste. *MSW Management*, Elements 2007, 32–40.

Ruch, Walter E., and Held, Bruce J. *Respiratory Protection*. Ann Arbor, MI: Ann Arbor Science Publishers, 1975.

Rudebeck, P. H., Buckley, M. J., Walton, M. E., and Rushworth, M. F. S. A role for the macaque anterior cingulate gyrus in social valuation. *Science* 313, September 1, 2006, 1311–12.

Rudolf, Michael P., Velders, Markwin P., Nieland, John D., et al. Vaccine design for DNA virus-induced cancer. In Peters, George A., and Peters, Barbara J. (Eds.), *Sourcebook on Asbestos Diseases*. Vol. 17. Charleston, VA: Reed Elsevier, 1998, 267–93.

Sentinel Event Alert, Wrong Site Surgery. Joint Commission on Accreditation of Healthcare Organizations, 2001.

Sentinel Event Alert, Tubing Misconnections. Oakbrook Terrace, IL: Joint Commission on Accreditation of Healthcare Organizations, issue 36, April 3, 2006.

Seward, Zachary M. Hygiene is shown to cut hospital blood infections. *Wall Street Journal*, December 28, 2006, D4. (Based on an article by Peter Pronovost of Johns Hopkins University that appeared in the December 28, 2006, issue of the *New England Journal of Medicine*.)

Sexton, J. B., Thomas, E. J., and Helmreich, R. L. Error, stress, and teamwork in medicine and aviation: Cross sectional surveys. *British Medical Journal* 320, 2000, 745–49.

Sheehan, Maura. Helping your community prepare POD's for mass prophylaxis for avian flu or bioterrorism. *The Synergist* 17(9), October 2006, 41–45.

Shojana, K. G., Dunkin, B. W., McDonald, K. M., et al. *Making Health Care Safer: A Critical Analysis of Patient Safety Practices*. Evidence Report 43. Washington, DC: Agency for Healthcare Research and Quality, 2001.

Six sigma cures hospital's error problems. *Quality Digest*, September 2006, 10.

Smith, Edward E., and Jonides, John. Storage and executive processes in the frontal lobes. *Science*, March 12, 1999, 1657–61.

Smith, Kip, and Hancock, Peter A. Situation awareness is adaptive, externally directed consciousness. *Human Factors* 37(1), 1995, 137–48.

SOX. *Sarbanes-Oxley Act of 2002*, 107th Congress of the United States, Washington, DC: GPO, January 23, 2003.

State Council AIDS Working Committee Office, U.N. Theme Group on HIV/AIDS in China. *A Joint Assessment of HIV/AIDS Prevention, Treatment, and Care in China*. Beijing, 2004.

Stoering, Petra. The impact of invisible stimuli. *Science* 314, December 15, 2006, 1694–95.

Summerfield, Christopher, Egner, Tobias, Greene, Matthew, Koechlin, Etienne, Mangels, Jennifer, and Hirsch, Joy. Predictive codes for forthcoming perception in the frontal cortex. *Science* 314, November 24, 2006, 1311–14.

Tapley, Asa L., Lurie, Peter, and Wolfe, Sidney M. Suboptimum use of FDA drug advisory committees. *The Lancet*, 368(9554), December 23, 2006–January 5, 2007, 2210.

Tesoriero, Heather Won. Texas jury's Vioxx finding stands, but plaintiff award lowered. *Wall Street Journal*, December 22, 2006, B4.

Thomas v. Hoffman-La Roche, Inc., 949 F.2d 806 at 812 (5th Cir. 1992).

Tinnerholm v. Parke-Davis & Co., 285 F. Supp 432, 411 F.2d 48 (2nd Cir. 1968).

Todd, Betsy. Emerging infections. *American Journal of Nursing* 106(9), September 2006, 34–35.

Too many hospital quality programs? *Quality Progress* 29(11), November 2006, 21.

Toshiba, Fujitsu join laptop-battery recalls. *Wall Street Journal*, October 1, 2006, A7.

Trials of War Criminals before the Nuremberg Military Tribunals under Control Council Law No. 10, Vol. 2, October 1946–April 1949. Washington, DC: Government Printing Office, 1947, 181–83.

Trussman, Susan. The shape of things to come? *American Journal of Nursing* 106(9), September 2006, 36–38. (Note: for the American Nurses Association position, see www.nursing world.org.)

Tryon, George H., and McKinnon, Gordon P. *Fire Protection Handbook*. Boston: National Fire Protection Association, 1969.

Tsushima, Yoshiaki, Sasaki, Yuka, and Watanabe, Takeo. Greater disruption due to failure of inhibitory control on an ambiguous distractor. *Science* 314, December 15, 2006, 1786–88.

Twenty Tips to Help Prevent Medical Errors. Patient Fact Sheet, AHRQ Publication No. 00-P038, September 2000.

U.S. Department of Defense. *MIL-STD-882D, Standard Practice for System Safety*. Washington, DC: February 10, 2000.

Van Cott, Harold P. and Kincade, Robert G. (Eds.). *Human Engineering Guide to Equipment Design*. New York: McGraw-Hill, 1972.

van Helden, Paul D., Victor, Tommie, and Warren, Robin M. The "source" of drug-resistant TB outbreaks. *Science* 314, October 20, 2006, 419–20.

van Rijswijk, Lia. So many dressings, so little information. *American Journal of Nursing* 106(12), December 2006, 66.

Vincent, C., Neale, G., and Woloshynowych, M. Adverse events in British hospitals: Preliminary record review. *British Medical Journal* 322, 2001, 517–19.

Vogel, Gretchen. Tracking Ebola's deadly march among wild apes. *Science* 314, December 8, 2006, 1522–23.

Wagner, Laura M., Capezuti, Elizabeth, Taylor, Jo A., Sattin, Richard, W., and Ouslander, Joseph G. Impact of a falls menu-driven incident-reporting system on documentation and quality improvement in nursing homes. *The Gerontologist* 45, 2005, 835–42.

Wagner, Peggy. "I'm sorry." *Minnesota Medicine*, November 2005, 24–42.

Watson, Donald, Crosbie, Michael J., and Callender, John Handock. *Time-Saver Standards for Architectural Design Data*. 7th ed. New York: McGraw-Hill, 1997.

Weinberg, Robert A. *The Biology of Cancer*. New York: Garland Sciences, 2006.

Weitzel, Tina, Robinson, Sherry, and Holmes, Jennifer. Preventing nosocomial pneumonia. *American Journal of Nursing* 106(9), September 2006, 72A– 72G.

Wenzel, Richard P., and Edmond, Michael B. Team-based prevention of catheter-related infections. *New England Journal of Medicine* 355(26), December 28, 2006, 2781–83.

Wessel, David, Wysocki, Bernard, and Martinez, Barbara. As health middlemen thrive, employers try to tame them. *Wall Street Journal*, December 29, 2006, pp. 1, A4.

Westphal, Silvia P. Pacemakers flagged in fresh recall. *Wall Street Journal*, June 27, 2006a, D-2.

Westphal, Silva P. FDA approves artificial heart sold by Abiomed. *Wall Street Journal*, September 6, 2006b, B4.

Whitehead, John W. Which is more dangerous to your health—the flu or the FDA? *Santa Monica Daily Press*, December 27, 2006, 4.

Whitlock, Jonathan R., Heynen, Arnold J., Shuler, Marshal G., and Bear, Mark F. Learning induces long-term potentiation in the hippocampus. *Science* 313, August 25, 2006, 1093–97.

Wise, Steven, P., and Desimone, Robert. Behavioral neurophysiology: Insights into seeing and grasping. *Science*, November 4, 1988, 736–40.

Woolf, Steven H., and Phillips, Robert L. A string of mistakes: The importance of cascade analysis in describing, counting, and preventing medical error. *Annals of Family Medicine*, July/August 2004, 317–326.

Wright, Alexi A., and Katz, Ingrid, T. Bar coding for patient safety. *New England Journal of Medicine* 353(4), July 28, 2005, 329–31.

Wu, Zunyou, Sun, Xinhua, Sullivan, Sheena, G., and Detels, Roger. HIV testing in China. *Science* 312, June 9, 2006, 1475–76.

Yoders, Jeff. Wall, clean thyself. *Building Design and Construction*, December 2006, 46–47.

Zoutman, Dick, Chan, Laurence, Watterson, James, Mackenzie, Thomas, and Djurfeldt, Marina. A Canadian survey of prophylactic antibiotic use among hip-fracture patients. *Infection Control and Hospital Epidemiology* 20(11), 1999, 752–55.

Appendix

The following lists illustrate the wealth of information contained in trade standards and guidance regulations. If medical products are distributed internationally, then the globally approved or harmonized standards should be used. The date of the standard is important because there are standards committees constantly attempting to update, expand, and revise each standard. Many earlier standards, pertaining to medical services, have been withdrawn. The standards are used for compliance purposes, which are minimal practice goals. Each standard has a consensus approval from those engaged in relevant activities. Some standards are specified contractually by original equipment manufacturers to their vendors or suppliers. The promulgating standards organization should be contacted for further information, such as the identity and availability of standards that are included by reference or listed in a primary standard.

Medical device manufacturers are first involved with the requirements of the Food and Drug Administration (FDA) contained in the Code of Federal Regulations (CFR), the relevant domestic trade standards, and the harmonized standards for world trade. They may have to register in countries where they actively market their product, conform to local requirements, have authorized local agents, and report adverse events under the vigilance systems then in effect. Standards compliance may be enforced with heavy penalties for noncompliance.

Standards Organizations

Abbreviations	Organization and Address
AAMI	Association for the Advancement of Medical Instrumentation 1110 N. Glebe Rd. Arlington, VA 22201-4795
ANSI	American National Standards Institute 11 W. 42nd St. New York, NY 10036
AS	Standards Australia 286 Sussex St. Sidney, NSW, Australia
CEN	European Committee for Standardization 36, rué de Stassart B-1050 Brussels, Belgium
CFR	Code of Federal Regulations U.S. Government Printing Office 732 N. Capitol St., NW Washington, DC 20401

Standards Organizations

Abbreviations	Organization and Address
CSA	Canadian Standards Association (or Organization) 178 Rexdale Blvd., Etobiocoke Ontario M9W 1R3 Canada
EIA	Electronic Industries Alliance 2500 Wilson Blvd. Arlington, VA 22201
EN	Harmonized European Standard Europe Union Rué de Genéve 1140 Brussels Belgium
IEC	International Electrotechnical Commission 3, rué de Varembé P.O. Box 131 CH-1211 Geneva 20 Switzerland
ISO	International Organization for Standardization 1, rué de Varembé Case Postale 56 CH-1211 Geneva 20 Switzerland
JIS	Japanese Standards Association 4-1-24 Akasaka Minato-ku Tokyo 107-8440 Japan
MIL-STD	Military Standard (United States) U.S. Government Printing Office 732 N. Capitol St. NW Washington, DC 20401
UL	Underwriters Laboratories, Inc. 333 Pfingsten Rd. Northbrook, IL 60062

Marks of Approval

CE	*Conformité Européenne*
	Product marking that indicates conformity with the health and safety requirements of the European Union
CC or CCC	*China Compulsory Certification*
	The safety approval process for products intended for use in the People's Republic of China

Selected Standards and Regulations

Function	Standard or Regulation
Agar Screening for Cytotoxicity	ASTM F-895-84 (2006)
Alarm Signals	BS EN 475:1995
Alarm Systems	IEC 60601-1-8:2003
Anesthetic Gas Monitors, Safety	ASTM F-1452-01
Auditing, Systems	ISO/TS 16949:2004
	ISO 19011
Bioabsorbable Plates and Screws	ASTM F-2502-05
Biocompatibility	ISO 10993-1:2003
Biocompatibility, Surgical Implants	ASTM F-981-04
Biomaterials, Absorbable	ASTM F-1983-99 (2003)
Bone Cement	ASTM F-451-99a
Bone Plates	ASTM F-382-99 (2003)
Bone Staples, Metallic	ASTM-564-02 (2006)
Breast Prostheses	ASTM F-703-96 (2002)
CE Mark Requirements Directive	98/79/EC
Clinical Investigation of Medical Devices for Human Subjects	ISO 14155
Diagnostic Devices, In Vitro	21 CFR Part 809. 10
	Directive 98/79 /EC:1998
	CEN/TC 140
DNA Repair Assay, Rats	ASTM E-1398-91 (2003)
Electrical Equipment, Electromagnetic Compatibility	IEC 60601-1:2004
Electrical Equipment Safety	AS/NZS 3200.1.2:1993
Electromagnetic System Effects	MIL-STD-464
Electromedical Current Limits	ANSI/AAMJ ES1:1998
Environmental Management	ISO 14001:2004
Electromedical Current Limits	ANSI/AAMI ES 1:1993
Environmental Management	ISO 14001:2004
Facial Implants	ASTM F -881-94 (2000)

Selected Standards and Regulations

Function	Standard or Regulation
Failure Mode, Effects, and Criticality	IEC 60812:1985
	MIL-STD-1629 A:1984
	SAE J-1739:1988
Fault Tree Analysis	IEC 61025:1990
Fixation Devices	ASTM F-1541-02
Fixation, Pins and Wires	ASTM F-366-04
Forceps, Terminology	ASTM F-921-85 (2002)
Hazard and Operability Studies	IEC 61882:2001
Hip Prostheses	ASTM F-2033-05
	ASTM F-1814-97a (2003)
Hip Wear Assessment	ASTM F-1714-96 (2002)
Human Factors Design Process	ANSI/AAMI HE 74:2001
Human Factors Usability	IEC 60601-1-1-6:2004
Implant Materials, Subcutaneous Screening	ASTM F-1408-97 (2002)
Informed Consent	21 CFR 50
Intermedullary Fixation, Care and Handling	ASTM 565-04
Intermedullary Fixation, Strength, and Stiffness	ASTM F-1264-03
Intermedullary Fixation, Surface Preparation	ASTM F-86-04
Knee Replacement	ASTM F-1223-05
	ASTM F-2083-06
	ASTM F-1814-97a (2003)
Labeling	21 CFR Part 801
	EN 375
	EN 376
	EN 591
	EN 592
	EN 980
	EN 1658
Labeling, Graphic Symbols	EM 980
Labeling, Home Use Instruments	EN 376
	EN 592
Labeling, Medical Device	FDA Guidance Document
Lasers and LEDs, Photobiological Safety	IEC 60825-1:2001
	CIE S 009/E:2002
Machine Production Performance	ISO/DIS 13700
Machine Safety	ISO 12100-1:2003
	ISO 12100-2:2003

Selected Standards and Regulations

Function	Standard or Regulation
Medical Device, Performance	EN 13532
Medical Device, Plates and Screws	21 CFR 888.3030
	ASTM F-543-02
Medical Device, Premarket Approval	21 CFR Part 814:2003
Medical Device, Quality Management	ISO 13485:2003
Medical Device, Registration and Listing	21 CFR Part 807
Medical Device, Reporting	21 CFR Part 803
Medical Electrical Equipment Requirements	IEC 60601-1 2004
	UL 60601-1:2003
	JIS T 0601-.1
	CAN/CSA No. 601.1 1995
	AS/NZS 32001.1:1998
	EN 60601-1
Medical Screwdriver Bits	ASTM F-116-00 (2004)
Misbranding (PMA)	21 CFR Part 807.97
Mutagenicity, Mouse Lymphoma Assay	ASTM E-1280-97 (2003)
Occupational Health and Safety Management	ANSI/AIHA Z10:2005
Orthopedic Fixation Devices, Angled	ASTM F-384-06
Orthopedic Implants, Care	ASTM F-565-04
Orthopedic Implant, Component Marking	ASTM F-983-86 (2005)
Osteosynthesis (fixation), Screw Corrosion	ASTM F-897-02
Oxygen Monitors	ISO 7767
Packaging Biodegradation	EN 13482
Packaging, Medical Devices	ISO 11607; 2006
	ANSI /ISO 11607:1997
	AAMI TIR No. 22:1997
	EN 868-1:1999
Packaging Recycling	EN 13430
	EN 13431
Packaging Reusability	EN 13429
Plastics, Extraction of Medical Materials	ASTM F-619-03
	ASTM D-1239
Processing Cross Contamination	21 CFR 1270.31(d)
Quality Acceptance Sampling and Testing	EN 13975
Quality Control	ANSI/ASQC B3:1996
	ISO/DIS 7870-1:1996
Quality Management Benefits	ISO 10014:2006

Selected Standards and Regulations

Function	Standard or Regulation
Quality Management Systems	EN ISO 13485:2000
	EN DIS 13485:2002
	EN ISO 13488:2000
	EN ISO 9001:2000
	EN ISO 9002:1994
	EN 9281:1995
	EN 29001
	EN 46001 Series
	EN ISO 928:1995
	EN 29001
	EN 46001 Series
Quality, Statistical Process Control	ISO 11462-1:2001
Quality System Regulations	21 CFR Part 820.30 and Part 11:2003
Radiation Control	21 CFR 1000-1050
Resuscitators, Human Safety	ASTM F-920-93 (1999)
Retrieval and Analysis of Medical Devices	ASTM F-561-05a
Risk Analysis	EN 1441
	ISO 14971:2000
Risk Assessment and Management	ISO 13485:2003
	ISO 14121:1999
	ISO 14971:2005
	EN 1441
	AS/NZS 4360:1999
Risk Assessment and Risk Reduction (Machine Tools)	ANSI B11.TR3
Risk Index (Overall)	ISO/IEC 14971:2000
Risk Management, Medical Devices	ISO 14971:2000
Safety Management, Occupational	ANSI/AIHA Z10:2005
	OHSAS 18001:1999
Shoulder Prostheses	ASTM F-1378-05
Silicone Elastomers	ASTM F-1027-86
Silicone Facial Implants	ASTM F-881-94 (2000)
Silicone Orofacial Devices	ASTM F-1027-86
Social Responsibility	ISO 26000 (draft version)
Software, Life Cycle	ISO /IEC 12207:1995
	IEEE 1074:1997
	AAMI SW 68:2001
Software Verification	IEEE 1012:1998
Special Controls	21 CFR Parts 800 to 898
Spinal Implants	ASTM F-1582 98 (2003)
	ASTM F-1717-04

Selected Standards and Regulations

Function	Standard or Regulation
Statistics for Rare Events	ASTM E-1263:1997 (2003) Section 12
Surgical Implant Inspection	ASTM F-601-03
Surgical Instrument Corrosion	ASTM F-1089-02
Surgical Stainless Steels	ASTM F-899-02
System Safety Design	MIL-STD-882 D:2000
System Safety (Navy)	SECNAVINST 5100.10 OPNAVINST 5100.20
System Safety Software	EIA SEB6-A:1990
System Validation	21 CFR 820.3(Z) 21 CFR 820.30 (f) (g)
Terminology, Polymeric Biomaterials	F-1251:89 (2003)
Tissue-Engineered Medical Products	ASTM F-2312-04
Transplantable Human Tissue	21 CFR Part 1270
White Blood Cell Morphology	ASTM F-2151-01
Wrist Implants	ASTM F-1357-99 (2004)

Subject Index

N

O

Z